Acknowledgements

This book does not pretend to be anything more than a very abbreviated account of the long and distinguished history of the 16th/5th The Queen's Royal Lancers. It could never have been written without the devoted work of previous historians of the Regiment and I gratefully acknowledge my debt to them. They were Major W. T. Wilcox, who wrote *The Historical Records of the Fifth (Royal Irish) Lancers*; Colonel Henry Graham, who wrote the *History of the Sixteenth (The Queen's) Light Dragoons (Lancers) 1759–1912* and *1912–25*; Brigadier C. N. Barclay, who wrote the *History of the 16th/5th The Queen's Royal Lancers 1925–61*; and Major H. G. Parkyn, who wrote a short *History of the 16th/5th Lancers*. I am also indebted to successive editors of the regimental magazine, *Scarlet and Green*. Most of the above histories are now out of print and the British Army has good cause to be grateful to Leo Cooper, the publisher of this Series, for making it possible for these short histories to be written, and for publishing them in such an attractive form.

It is impossible to cover every aspect of a regiment's history in the space of 30,000 words and therefore much of interest has to be omitted. It is a matter of judgement whether to omit or include this or that episode and no two readers are likely to agree on this score. Some may complain that too much space has been devoted to the distant past and not enough to events which have occurred within the memory of those still living. I would agree that this is legitimate criticism but would point out that the strength of the British Army's regimental system lies in the fact that its roots are so firmly embedded in the past. This has enabled it to undergo periodic reorganizations and emerge successfully from all of them. Unless those who are serving in the Regiment today have some idea of what happened in the past, they cannot be expected to understand the tradition which binds us together in comradeship. I make no excuse for devoting more space to the earlier years than I have done to the recent past, but those who wish to know more about the part the Regiment played in the two great wars of this century can confidently be recommended to Graham's and Barclay's accounts of those wars.

General Sir Hubert Gough defined *cavalry spirit* as a combination of independence of thought, quickness in decision, and boldness in action. Those who fight mounted, be it on horse or in tank, must be able to seize the fleeting opportunity which may never recur. When the machine-gun and barbed wire ousted the horse from the battlefield, many soldiers of many nations believed that cavalrymen would never be able to adjust themselves to a new kind of mount. There were also many cavalry officers, some of great distinction, who continued to argue for the retention of the horse long after the argument had lost all validity. Both schools of thought were proved wrong, as this short history will show. The 16th/5th The Queen's Royal Lancers, in common with their sister cavalry regiments, proved convincingly between 1939 and 1945 that they could master the new form of mobile warfare without in any way losing the *panache* and *élan* which has traditionally been associated with the horsed soldier.

It was the custom in the 16th Lancers for much of their long history that when the Commanding Officer had occasion to address the Regiment he would preface his remarks with the words, 'Gentlemen of the Lancers,' and this at a time in British history when the soldier was regarded as coming from the dregs of society. Those times are happily past and forgotten but I, in my turn, must thank the 'Gentlemen of the Lancers' for making it possible for me to write this history, and for making my soldiering days with them so memorable and happy. I must also thank the Colonel of the Regiment, Colonel Anthony Bullivant, for entrusting me with the writing of this history; the Commanding Officer, Lieutenant-Colonel Alastair Dennis, and Major Tony Rundell for their help in the selection of illustrations; and Lieutenant-Colonel John Pownall who kindly read the manuscript and provided me with useful advice. Finally I must thank Miss Margaret Phillips who has so capably typed the manuscript and coped with my numerous deletions and alterations.

J.D.L.

Abinger Hammer.
17 March, 1972.

INTRODUCTION
Lt.-General Sir Brian Horrocks

AN Introduction should always be an appetiser for a good literary meal, and I only hope that this one will encourage people to read the author's fascinating account of two famous Cavalry Regiments which were amalgamated in April, 1922.

General Lunt not only writes extremely well and is therefore easy to read but he has also had an unusually wide experience of the British Army. He has served in both the Infantry and the Cavalry and ended his career as Vice Adjutant General. Now that he has retired I hope he will write many more books of this high calibre.

This book is concerned with the story of the 5th Royal Irish Dragoons and the 16th Lancers. The 5th were formed on 1 January, 1689, and their first major engagement was at the Battle of the Boyne on 14 June, 1690, when William III went to Ireland and defeated James II. Whereas many far more important battles have long since been forgotten, The Boyne is still remembered with concentrated bitterness in the south of Ireland and joyfully in the north.

The many battles in which the 5th have been engaged have all been dealt with most adequately by the author, so I shall only dwell on those operations which have vitally affected their history.

The Regiment's finest hour was unquestionably when they were fighting in the Low Countries during the War of the Spanish Succession, while serving under the finest General this country has ever produced, John Churchill, subsequently Duke of Marlborough. His famous 600-mile march was aptly described by Sir Winston Churchill, as 'The Red Caterpillar which crawled across Europe'. Marlborough not only deceived his Dutch Allies as to his ultimate destination, who would otherwise never have allowed him to go, but through a highly organized Intelligence Service, he also knew exactly what the French were doing.

Marlborough really cared for the welfare of his troops and was the first British General to take pains over administration. On arrival at their different camps which had all been carefully reconnoitered beforehand, the soldiers would find them stocked with all that they required. This would be taken for granted today, but in the early eighteenth Century 'man management', as it is now called, was unheard of. The result was that the moral and physical wellbeing of his men at the end of the march was higher than when it started.

The 5th were in the thick of the Battle of Blenheim which smashed the French military dominance of Europe. When the battle was won, Marlborough, who had been in the saddle all day, called for a drum, which was handed to him by a trooper of the 5th. He searched his pockets but could only find a 'bill of fare' and on the drum he wrote the historic letter, which is still kept at Blenheim, asking Sarah, his wife, to inform Queen Anne that 'her Army has had a glorious victory, etc.' The 5th also distinguished themselves at Ramillies, Oudenarde and Malplaquet and by the end of the war were a really well-disciplined fighting unit. On their return to the United Kingdom they were, alas, posted once more to Ireland, where they remained for the next 86 years. This was probably the lowest ebb to which the Regiment ever sunk; there was practically no contact at all between the officers and their men. Most of the former had paid up to 500 guineas for their commissions (a lot of money in those days) and although they were perfectly prepared to charge gallantly into battle, they found garrison duties in Ireland so uninteresting that 2/3rds of them were away on leave for most of the year. The only time the Regiment was concentrated was for the Annual Review. As the author says,

> 'The inevitable consequence of such service was absentee officers and bored, demoralised soldiers; mutiny was not infrequent and insubordination was common'.

After the Rebellion of 1798 the Lord Lieutenant recommended that the Regiment should be removed from Ireland, and on 8 April,

King George III ordered its disbandment – and so ended the first chapter in the life of the 5th Royal Irish Dragoons.

Let us now turn to their present partner. On 4 August, 1759 it was decided to follow the example of certain foreign armies and form some regiments of light cavalry. John Burgoyne was ordered to raise the 16th Light Dragoons which he did without difficulty. The 16th were fortunate because, although discipline throughout the army was so brutal that in 1770 the Americans called the British soldiers 'Bloody Backs' on account of the constant use of the lash, Burgoyne had very strong views on the relationship which should exist between officers and men.

He emphasised the need to treat soldiers as thinking beings and insisted on his officers getting insight into the character of each man. Initially he addressed his officers as follows: –

> 'There are occasions such as doing stable or fatigue duty when officers may slacken the reins so far as to talk with soldiers, nay, even a joke may be used, not only without harm but to good purpose, for condescension well applied are an encouragement to the well disposed, and at the same time a tacit reproof to others'.

I am glad the author has included this passage because, strange as it may seem today, in the middle of the seventeenth Century it was radical stuff indeed. But herein lies the core of the many successes which have attended this Regiment: it really was a home for the soldier of those days, who had little enough to live for anyhow. Unfortunately, the cavalry officers as a rule consisted largely of rich young men who spent most of their time on leave and probably did not know even the names of 50% of the men under their command. Their life consisted of Balls which they attended in the most gorgeous uniforms imaginable, much to the delight of the female sex, and fox hunting, of course. Their one and only asset was their bravery.

In the Infantry, on the other hand, the officers were far from rich and, as their Regiments, unlike the Cavalry, were called upon to provide garrisons all over the world, they were bound to get to know and look after their men, and this tradition has persisted. When I was commanding a platoon in 1914 woe betide me if I attempted to have my own meal without first reporting to my Company Commander 'No. 10 Platoon settled into their billets and fed, sir'.

This was the spirit which Burgoyne implanted into the 16th, and is the reason why the Regiment has always been popular with other units as well as being one of the most efficient cavalry regiments I have ever met.

The 16th went from strength to strength and on 20 May, 1766, after an Inspection on Wimbledon Common by the Commander in Chief, King Charles III, he commanded that their title should be 'Queen Charlotte's Light Dragoons'; ever since they have been called 'The Queens'. The King also took great interest in the dress of his troops and ordered that the 16th should wear scarlet coats with blue facing.

From then on they appear to have taken part in every major war in which Britain has been involved with the exception of the Crimean War. For three years they fought in the American War of Independence. On one occasion, under the command of William Harcourt, who later became Field Marshal Earl Harcourt and Colonel of the Regiment, they captured the famous American General, Charles Lee, who some fourteen years previously had been their own C.O. in Portugal.

After returning to England they were unfortunate enough to take part in the worst organised Campaign in which the British have ever been engaged – the War against the Revolutionary Armies of France. It was here that the British Commander, the Noble Duke of York, marched his men to the top of the hill and marched them down again. It was a truly inglorious campaign, notable in one respect only – it was here that Arthur Wellesley, subsequently Duke of Wellington, saw active service for the first time, and distinguished himself in command of the rearguard. When asked whether he had learned anything from this Campaign he replied, 'Why – I learned what one ought not to do, and that is always something'.

Their next Campaign came in April 1809, when they joined Sir Arthur Wellesley, in the Peninsular, and did not return to England for 5 years, in the course of which time they were present at 7 pitched battles and lost 300 officers, N.C.Os and men, plus 1,416 horses. Thanks to their previous training they distinguished themselves particularly on outpost duty. They even impressed Wellington, who regarded the average cavalry officer as an amateur. The 16th, on the other hand, became noted for their control on the battlefield. At Waterloo, they charged the French as part of Vandeleur's Brigade, in order to extricate Ponsonby's heavy cavalry which had got out of control, an occurrence which happened all too often with the British Cavalry of those days.

After Napoleon had finally been beaten the 16th returned to the United Kingdom but in 1816 they went back to Ireland. It was at this period that they were equipped with lances and their title was changed to 'The 16th the Queen's Lancers', nicknamed the 'Scarlet Lancers'. Wellington did not approve of lances, but he had not been present at the Battle of Albuhera when the Polish Lancers caught a British Brigade on its way to the threatened right flank and caused them terrible casualties. It is an astonishing fact that lances were still carried in some Horse Cavalry Regiments in the Mounties Division which was dispatched to the Middle East at the start of the Second World War.

In 1822 the Regiment was sent to India, where it remained for 24 years and took part in the invasion of Afghanistan. It is very difficult for us in these days of mechanization to imagine what an Army in the field in India looked like in the middle of the nineteenth century. The 16th had nearly 5,000 followers to administer to its needs, and they were only one of several regiments. They also took a pack of fox hounds with which to while away the dull moments. This vast mass of men, animals, carts, etc., marched 1500 miles across desert and barren mountains and, on arrival in Kabul, one of their first acts was to lay out a race course and cricket pitch; their return to India took nearly four months' steady marching: they covered 2,483 miles in 463 days.

During the first Sikh War the Regiment saved the day at Aliwal on 28 January, 1846. Though outnumbered many times over, they defeated the Sikhs, but their casualties were heavy and by the time the Cease Fire sounded they had lost at least 1/3rd of their strength. The Charge at Aliwal is celebrated by the Regiment each year and the author quite rightly devotes a whole chapter to a vivid description of the battle.

The 16th have always been famous for their Commanding Officers, and in my opinion the greatest of them all was Hubert de la Poer Gough. Gough came from a famous Anglo-Irish family. His father, his uncle and his brother had all won the V.C.

Another famous character to serve with the Regiment was William Robertson, son of a village postman, who rose rapidly through the ranks to become Troop Sergeant Major. He was commissioned into the 3rd Dragoon Guards in 1888, and although he had no private means at all by 1910 he was a Major General and Commandant of the Staff College. In 1915 he became C.I.G.S. and ended his career as a Field Marshal and a Baronet. His son, Lord Robertson of Oakridge, had an almost equally distinguished career: he was Montgomery's very able Senior Administrative Staff Officer during the Desert Campaign and ended his military career as Adjutant General. The presence of men like these had a profound effect on the young officers of the 16th Lancers.

On 9 January, 1858, Queen Victoria commanded that the 5th Royal Irish Dragoons, who had been disbanded 60 years before, should be resuscitated and very soon they became respected as a most efficient regiment. At the end of some manoeuvres in India their C.O. wheeled the Regiment into line, ordered the Officers' Call to be sounded and, after they had galloped up to him, he said, 'Gentlemen, I have called you out to look at such a Regiment of Cavalry as you are unlikely ever to see again, turn about and look at the Regiment'.

Both the 16th and 5th benefitted from their experiences in South Africa, and as a result were in much better shape than some of the others to cope with this new concept of warfare. The standard of C.O. continued to be very high. In 1902 the 5th had Lt.-Colonel E. H. Allenby – known to everyone as The Bull. I first met this huge, imposing figure in August, 1914, when, in command of the first reinforcements for my Regiment which was then serving with the B.E.F. I landed at St. Nazaire.

The First World War was a nightmare for the Cavalry. The 16th and 5th covered the Retreat of the B.E.F. from Mons in one of the few mobile operations which occurred during the whole war; both then took their turn as infantry in the trenches.

It would be wrong to pass over this terrible slaughter without mentioning Gough who, at 46 years of age, was given command of the 5th Army. His army was under-manned and over-stretched, a fact which he pointed out continuously without result. It was his Army which bore the brunt of the German offensive in March, 1918. Gough fought an able delaying action and in the end stopped the Germans, but a scapegoat had to be found. He was sacked without hope of redress on 28 March, 1918, and retired from the Army in 1920, an embittered man. Fortunately he was completely exonerated when the official history was published in 1936 and 1937. He was made a G.C.B. and was Colonel of the Regiment from 1936 to 1943. He will always be remembered as one of the greatest and most attractive officers to serve in this famous Regiment.

As after all great wars the armed forces were reduced to an almost dangerous level, and in the Cavalry this was done mainly by amalgamation, so in April, 1922, the 16th and 5th were combined into one Regiment, the 16th/5th. At first they did not prove happy bedfellows, which was hardly surprising as each was a highly individual unit. It was a difficult time for them both because mechanisation now reared its head. To every thinking soldier it was obvious that the horse had no place on the modern battlefield. But as the lives of most cavalry regiments had always centred round the horse, the changeover very nearly broke their hearts. In 1940 the 16th/5th received their first tanks – ancient Crusaders and Valentines. Thanks to Lt.-Colonel Macintyre, the Regiment gradually absorbed these new arms. As a wise old General once said to me, 'One good Commanding Officer is worth a ton of tradition with mould on it', and this was certainly true of

Macintyre and the 16th/5th.

When they first came under my command in North Africa they were equipped with the latest Sherman tanks, and there was no better armoured regiment in the British Army. I had been sent from the 8th Army to the 1st, to command 9 Corps for the final attack to capture Tunis. The famous Desert Rats (7th Armoured Division) accompanied me from the 8th Army and, in addition, General Anderson gave me the 6th Armoured Division, in which the 16th/5th were serving. After the 4th British and 4th Indian Infantry Divisions had punched a hole in the German defences, I launched these two Armoured Divisions from Medjez-el-Bab up the valley, via Massicault, straight on to Tunis. It was a wonderful sight to see these two well-trained armoured divisions being used in a role for which they had been specifically designed, deep penetration of an enemy position. If they encountered the enemy holding a defensive position their guns would drop into action, their Infantry would debus and attack while the tanks outflanked the enemy positions. The whole operation was, of course, controlled by wireless. This sounds easy, but it needs a high standard of training which both these Divisions possessed. When it became obvious that the enemy resistance was broken, I ordered Charles Keightley to turn his 6th Armoured Division to the south east through Hammam Lif to Hammamet, in order to cut off the Germans who had escaped into the tip of the Cap Bon Peninsular. Throughout the night, the 6th Armoured continued to advance. By next morning Tunis was captured and the Germans in Cap Bon sealed off. The war in Africa was over, and the last troops to fight there were the 6th Armoured Division. By 13 May some 217,600 German and Italians had surrendered.

This was the last time I saw them because a few days later I was wounded and evacuated to the United Kingdom.

For the 16th/5th the rest of the War consisted of a hard slog up Italy, mainly in mountains and mud, not suited to armoured warfare, but as might be expected their morale remained as high as ever. In the author's words, 'Of their many battle honours Italy 1944-45 was harder earned than most'.

Since the War, they have been employed in a peace keeping role all over the world, often split up into squadrons, and equipped with armoured cars. It is a curious fact that they are now once again in Northern Ireland, where they first saw the light of day some 300 years ago.

Anyhow, wherever they have served, they have always been welcomed, because they are not only highly professional but also friendly, and entirely without snobbery of any sort.

16th/5th The Queen's Royal Lancers

Chapter
I
The Royal Dragoons of Ireland

IT is said that the Englishman who understands the Irish has yet to be born, but on the other hand, as George Meredith wrote in *Diana of the Crossways*, ' 'Tis Ireland gives England her soldiers, her Generals too.' The truth of this can be seen in the cemeteries across the world where lie the Irish soldiers and the Anglo-Irish squirearchy who led them. The army that won for Wellington the campaign in Spain and Portugal was full of Irishmen, and at Waterloo, according to Kincaid, 'The 27th Regiment (Inniskilling) were lying literally dead in square'. It cannot be denied that Irishmen played an important part in conquering the British Empire and in defending it once it was won. Many of the regiments in whose ranks they fought have long since been disbanded and only the memory of them remains. Few had such a chequered career as the cavalry regiment raised towards the end of 1688 on the orders of Gustavus Hamilton, Governor of Enniskillen, which later became the Royal Dragoons of Ireland, and later still the 5th Royal Irish Lancers.

It is a melancholy reminder of the 'Irish Problem' that 300 years ago Protestants and Catholics stood face to face, as they do today, at the gates of Londonderry. King James II, recently ousted from his throne, had landed in Ireland and placed his fate in the hands of the Earl of Tyrconnel. The Protestants in the north rallied under Gustavus Hamilton to defend the cause of King William III and two regiments of Dragoons and three of Foot were hastily raised in Enniskillen. In July, 1689, General Kirk arrived from England to relieve Londonderry and on the 20th of that month issued commissions to the officers of the Enniskillen regiments in the name of King William III, who subsequently ratified them. Captain James Wynne, 'a gentleman of Ireland, but then a captain in Colonel Stuart's

Regiment', was appointed to be Colonel of the Dragoons with the pay of 15 shillings a day, and according to the then prevailing practice his regiment took its title from his name and was known as 'Wynne's Inniskilling Dragoons'. Together with the other Enniskillen regiments, or more properly levies, it was brought on to the establishment of the Regular Army by a Royal Warrant dated 'the First day of January, 1689, in the first year of our Reign.'

It was not an efficient unit, even when judged by the not very exacting standards of the time, and has been compared to a mob of undisciplined boys led by officers who were ignorant, negligent and useless, but they could fight well enough. Dragoons were essentially soldiers who used their horses to convey them to a suitable tactical position on the battlefield, whereupon they usually dismounted and continued the fight on foot. Horses were in short supply in Ireland, and the transport arrangements across St George's Channel from England were so scandalously conducted that one regiment lost every horse in the passage. It was fortunate for William III that Tyrconnel's men were no better trained nor equipped, and that there was time to lick the raw Enniskillen levies into shape.

This was the task of the Duke of Schomberg, a tough old German who arrived in Ireland in August, 1689 to command the King's forces. By then Wynne's regiment had been blooded at the battle of Newton Butler towards the end of July, when we are told 'the name of the Enniskillen men became a terror to the Irish'. Schomberg was much concerned by 'the horrid and detestable crimes of profane cursing, swearing, and taking God's Holy Name in vain'. This moved him to issue a special order of the day on the subject, dated 18 January, 1690, charging and commanding all officers and soldiers to 'forbear all vain cursing, etc., etc.'

William III went to Ireland on 14 June, 1690, and shortly afterwards defeated James II at the battle of the Boyne.

The battle took place on 1 July and began inauspiciously for King William by the death of old Schomberg. Tradition has it that he was shot by an Irishman with his duck gun as Schomberg rode down to reconnoitre the river bank; over 200 years later a monument was erected close to his grave in St Patrick's Cathedral in Dublin by the

5th Royal Irish Lancers, descendants of Wynne's Dragoons, to commemorate those of their comrades who had fallen in the South African War. The battle was bitterly contested and at one stage King William placed himself at the head of the Dragoons from Enniskillen, saying, 'Gentlemen, I have heard much of your exploits, and now I shall witness them.' The English cavalry then crossed the river and charged with such fury that they scattered the enemy infantry but then galloped off out of control. Fortunately for King William the Irish infantry broke at the crucial moment and took to their heels; only the gallantry of the Irish horse, who charged repeatedly, saved their infantry from massacre, but even so the pursuit was a bloody business.

Later Wynne's Dragoons took part in the unsuccessful sieges of Limerick and Athlone and there is a sadly contemporary touch in a description of the assault at Limerick during August 1690 – '. . . for three hours did a sharp fight continue, in which the Irish women boldly joined; and when they failed to obtain more deadly missiles, threw stones and broken bottles.' Wynne's Dragoons were also present at the battle of Aughrim on 12 July, 1691, which virtually concluded the war and where, according to an eye witness, Captain Parker, the Irish never fought so well in their own country as they did on that day; but all their gallantry was in vain.

In May, 1694, Wynne's Dragoons were sent to Flanders to join the Allied armies collected to fight the French. Wynne's regiment was ill-found so far as horses were concerned, since most of them were sadly out of condition after crossing the Irish Sea and English Channel. There was, however, time enough to improve on this since warfare at that time can hardly be described as fast-moving. In fact, it was more a matter of sieges and patrol actions than set-piece battles, and when the latter did occur it was all according to regulations and extremely formalized. Nevertheless men were killed and men were wounded, most of the latter dying, since medical attention was mainly confined to amputations and to bleeding. Wynne was wounded in the knee at Moorsleede, where his Dragoons defended a convoy of supplies against enemy attack. The wound was slight but Wynne died shortly afterwards, and William III gave the Colonelcy to Charles Ross, who had served as one of his Aides-de-Camp. In July, 1695, Wynne's became Ross's Dragoons.

The war in the Low Countries ended in 1698. Ross's Dragoons lost far more men from disease than from enemy action and like every other regiment they had many deserters. The wonder of it is that men could be induced to serve under such wretched conditions, with inefficient officers, scoundrelly commissaries, and doctors who made better butchers than surgeons. But for men who came from the rural areas of Ireland life on campaign in Flanders differed little from the hardships of life in a Connemara cabin, and there was always the prospect of loot. Ross's Dragoons returned to Ireland where the garrison was fixed at 12,000 men. This led to a reduction in the strength of regiments. Ross's Dragoons were to consist of eight Troops, amounting to 362 officers and men, and these were billeted in inns and lodging-houses throughout Connaught. Maintaining discipline among these scattered detachments was always a problem in Ireland, where the army operated as a gendarmerie, and although life on the whole was easy-going, punishments when ordered could be draconian. Six of Ross's Dragoons were sentenced by Court Martial in December, 1698 'to run and be whipped three several times by an entire regiment of Foot drawn out for that purpose on three several days on St Stephen's Green.' The punishment known as the Gatloup was carried out by troops paraded in open ranks. Each man carried a stout stick. The ranks were faced inwards and the prisoner, stripped to the waist, was marched up and down the lines of men. Each man was expected to strike him with maximum force on the 'naked back, breast, arms, or where his cudgel should light,' while the screams of the victim were intended to be drowned by the drums which beat throughout the punishment. Many men died while undergoing this savage ordeal.

In March, 1702, Ross's Dragoons were dispatched to join Marlborough's army in the Low Countries. They were therefore present at Blenheim where they were in the thick of the fight. Marlborough's pencilled note to his wife, written in the fading light after a long and exhausting day, must surely be one of the shortest dispatches in history:

13 August, 1704.
I have not time to say more, but to beg you will give my duty to the Queen, and let her know her army has had a glorious victory. Monsr. Tallard and two other generals are in my coach, and I am following the rest. The bearer, my aide-de-camp, Colonel Parke, will give her an account of what has pass'd. I shall do it in a day or two by another more at large.

Ross had petitioned that his regiment should be known as The Royal Dragoons of Ireland and in March, 1704, that title was formally conferred on the Regiment. It was therefore as the Royal Dragoons of Ireland that the regiment charged at Blenheim, capturing three French kettle-drums which Marlborough directed should henceforward be carried at the head of the regiment. These kettle-drums are believed to be still in existence, in the Queen's Armoury in the Tower of London, and on 19 March, 1959, when the Regiment received the Guidon from the Queen at Buckingham Palace, the Blenheim kettle-drums were brought from the Tower and piled in front of the parade. On them was laid the Guidon before its consecration, 255 years after their capture at Blenheim.

In 1706 the Royal Irish Dragoons, brigaded with the Scots Dragoons (Scots Greys), distinguished themselves at the battle of Ramillies on 23 May. The two regiments took two battalions of the *Régiment de Picardie* prisoner and destroyed the third battalion. They were rewarded by the distinction of being allowed to wear Grenadier caps, thereby differentiating them from other cavalry regiments. This kind of distinction, jealously regarded, was a cheap and effective way of making up for the disadvantages of a soldier's life. The Royal Dragoons of Ireland, campaigning in the Low Countries under Marlborough, did not find themselves all that much better off materially than Ross's Dragoons had been under William III, except that Marlborough did give them victories and organized a better supply system. Morale was good, despite peculation and nepotism. Marlborough, writing on 19 December, 1706, to the Earl of Cardigan, who had requested that the son of the late Major-General Brudenell should be given a Company, delivered only the mildest rebuke to what he must have considered an outrageous suggestion:

'I have so just a sense of the father's good services that I shall always be glad to embrace any opportunity of showing it to his family; but your Lordship tells me he is not above five years old'!

The Royal Dragoons of Ireland were present at Marlborough's other two great victories, Oudenarde on 11 July, 1708, and Malplaquet on 11 September, 1709. They had also helped to capture Bruges in 1706 and took part in what many regard as the most brilliant of all Marlborough's operations – the passage of the French lines on 4 August, 1711. It is sad that this brilliant feat should have been followed by Marlborough's recall and subsequent disgrace. Few British commanders have been loved so well by their soldiers as Marlborough.

Charles Ross, now a Lieutenant-General of Horse, took some part in the negotiations which concluded the war, and in June, 1713, he was informed that 'the Regiment under your command is to be put on the Establishment of Ireland and to be paid for by the revenues of that country'. Three months later Ross was appointed Envoy Extraordinary in Paris. It was to be the fate of his regiment to serve continuously in Ireland for the next 86 years. The 5th Royal Irish Dragoons, as they are shown in the 1752 Army List, were never given the opportunity to escape from the dreary round of garrison duties which were the lot of a regiment on the Irish establishment. They hunted down smugglers, galloped after highwaymen, held down a discontented and impoverished peasantry, and annually came together from their outlying detachments to take part in a Review.

'Cavalry Corps in Ireland were extremely select,' wrote Surgeon John Smet of the 8th Light Dragoons in 1784, 'as from the very low establishment, it was in the power of the Colonels of choosing among a number of young gentlemen of distinction who might wish to get a commission, and who all could easily afford to add a hundred pounds a year to their pay. The warrants were also purchased at a high price, often by the sons of gentlemen for as much as five hundred guineas. The privates were always young men well recommended and whose connections were known. Indeed, the dragoon service was at that time extremely easy and pleasant, so much so, that when a vacancy happened, several desirable recruits always offered, and the men selected in general, got no more than one shilling bounty.'*

Pleasant it may have been but it was certainly not soldiering. Smet tells us that two-thirds of the officers were away for most of the year on leave, and the men did not do too badly either. They took their horses home with them and seldom wore uniform. To save the expense of forage horses were put out to grass for as long as possible and must in consequence have been unfit for hard work for most of the year. 'Such a service had many attractions,' says Smet. The annual review brought everyone together again, and, 'The Officers now meeting again, after such a long separation from each other, in affluent circumstances, which they had improved while they had lived with their friends, justly looked on the time of the year they were to be reviewed in as the pleasantest season. The mornings were spent at exercise and the remainder of the time in festivity.'

There were problems, of course, since it was not always possible to find a suitable field for a review. James Stewart, Lt-Col 5th Dragoons, drew attention to the fact in a letter dated 16 June, 1785, from Clonmel:

'That your Memorialist took a field of exercise for the 5th Dragoons during their assembly and review at Cashell, in the last and present months, as there was no common ground near that town. That he was obliged to pay 13 guineas for that field, as he could not get a proper place for less money. Praying to be allowed to charge the said sum to the contingent bill of said Regiment in the usual manner.'

Ireland was in a sorry state towards the end of the eighteenth century. There was much poverty and great unrest – if there had not been there would have been no need to station detachments of the 5th Royal Irish Dragoons round the countryside to keep the peasantry in order. The inevitable conseqences of such service were absentee officers and bored and demoralized soldiers. Mutiny was not infrequent and insubordination was common. General Abercromby was appalled when he arrived in Ireland in 1798 as Commander-in-Chief. 'The very disgraceful frequency of courts-martial and the many

* Smet's *Historical Record of the 8th Hussars.*

complaints of irregularities in the conduct of the troops in the Kingdom, have too unfortunately proved the army to be in a state of licentiousness which must render it formidable to every one but the enemy,' stated his first General Order. He promptly resigned his appointment and handed over to his deputy, General Lake, who was later to find India a more rewarding place than Ireland to win distinction.

If regiments recently arrived in Ireland succumbed so quickly to the general malaise, it is hardly surprising that regiments like the 5th Royal Irish Dragoons, which had known no other garrison since Marlborough's campaigns, should have been in such a sorry state. Their inspection report in 1797 stated, 'A very essential change is absolutely necessary to put the 5th Regiment of Dragoons in a condition for service which at present they are entirely unfit for,' but nothing was done. On 23 May, 1798, a rising took place in Ireland and detachments of the 5th Royal Irish Dragoons found themselves fighting for their lives. Like all rebellions of its kind the Irish Rebellion of 1798 was marked by cruelties on both sides. There were various engagements, the best known of which is Vinegar Hill, but by September the English had regained the upper hand. The 'Irish Problem' was once again settled – for the time being at least.

The regiments on the Irish establishment had been found deficient in the essential military virtues. Some even seemed to be deficient in loyalty, such as the 5th Royal Irish Dragoons whose ranks apparently contained certain men who had enlisted with the intent of joining the rebels and seducing other soldiers from their loyalty. These were men who had taken the oath of the United Irishmen and some of them were discovered and sentenced to death by court martial. At this crucial moment the commanding officer was absent on leave. The Lord Lieutenant recommended the regiment's removal from Ireland to England where it could be brought back into a state of efficiency, and its Fenian element weeded out, but King George III thought differently. In an order from the Horse Guards dated 8 April, 1799, it was stated that the King had decided to disband forthwith his 5th, or Royal Irish Regiment of Dragoons.

A report on the regiment's last days comments favourably on its good behaviour while awaiting disbandment. 'We cannot conclude,' says the report, 'without expressing our regret that a Regiment which has so frequently deserved well of its country, should have incurred His Majesty's displeasure. Let us hope, that like the Phoenix, it may some time or other rise out of its own ashes, be restored to the Army, and add fresh laurels to those of Ramillies and Hochstet.'*

On 10 April, 1799, at Chatham, the 5th or Royal Irish Dragoons were disbanded, and its officers, N.C.O.s and soldiers dispersed to other regiments.

---

* Hochstet—Blenheim.

Chapter 2
Burgoyne's Light Horse

DURING the wars between Frederick the Great and the Empress of Austria a new kind of cavalry appeared on the battlefields of Silesia. They came to be known as Hussars, a Hungarian word by derivation, since the horsemen were mainly recruited from the great plains of Hungary, and they were used almost exclusively for scouting and outpost duties. They were very lightly equipped and mounted on ponies which could live off the country and make up in speed and endurance for what they lacked in looks. It was not long before other armies began to copy the Austrian organization and, as a result, there evolved three distinct types of cavalry – heavy cavalry for shock action, light cavalry for reconnaissance, and a form of mounted infantryman who most approximated to the cavalry soldier of the preceding two hundred years. In the latter case horses were merely a means of transporting the soldier to the scene of action where thereafter he fought on foot, and cavalry of this type were usually known as Dragoons.

In the year 1759 the British decided to follow the example of the Austrian and Prussian armies and raise some regiments of light cavalry which were to be known as Light Dragoons. On 17 March, 1759, orders were issued for the raising of the 15th Light Dragoons, and four months later similar orders were issued for the raising of the 16th Light Dragoons. The officer chosen by King George II to raise the latter regiment was a certain John Burgoyne of the 2nd Foot Guards. He is known chiefly, if not entirely, for his surrender to superior American forces at Saratoga on 16 October, 1777, and like most unsuccessful British generals he has been written off as incompetent. Burgoyne was, in fact, an extremely able soldier and a man of considerable parts and much intelligence.

He was born in London in 1722. His father was the second son of a Bedfordshire baronet and he lived in fine style but died in a debtor's prison. Horace Walpole, that notorious scandalmonger, spread a rumour in later years that John Burgoyne was in fact the natural son of Lord Bingley, at one time ambassador in Spain, who was Burgoyne's godfather and who left his mother a considerable legacy, but research has failed to substantiate this charge. He was educated at Westminster School where he became friendly with Lord Strange, son of the Earl of Derby, and at some stage in this friendship became acquainted with Strange's sister, Lady Charlotte Stanley. Acquaintanceship soon ripened into something more and in 1751, when he was twenty-nine, Burgoyne eloped with the daughter of one of the most influential noblemen in England. At the time of his elopement Burgoyne was serving in the 1st Royal Dragoons, in which he had purchased in rapid succession, a Lieutenancy and a Captaincy; but the Derby family disapproved of the match and Burgoyne and his wife were forced to escape their creditors by fleeing to France. In 1756 he re-entered the army, and purchased a Captaincy in the 11th Dragoons. Soon after this, war broke out with France and Burgoyne distinguished himself in the operations against St Malo and Cherbourg.

Already well known in London society for his good looks, his social graces, his daring as a gambler, and his connection with the Derby family, Burgoyne's name soon became known at Court whence all worthwhile military patronage flowed. It was not long before he was transferred as a Captain and Lieutenant-Colonel into the 2nd Foot Guards. He was 36. When it was decided to raise two regiments of Light Cavalry the choice of Burgoyne to command one of them must have seemed natural. On 4 August, 1759, he was appointed Lieutenant-Colonel Commandant of the 16th Light Dragoons and directed to raise four troops of light cavalry in the general area of Northampton. There seems to have been no problem in obtaining either recruits or officers and there were no less than three Honourables and three Baronets on the regimental list within six months of the 16th's formation.

He began by issuing a recruiting poster which he must have

drafted himself since it bears all the evidence of his pen:

'You will be mounted on the finest horses in the world,' it ran, 'with superb clothing and the richest accoutrements; your pay and privileges are equal to two guineas a week; you are everywhere respected; your society is courted; you are admired by the Fair, which, together with the chance of getting switched to a buxom widow, or of brushing with a rich heiress, renders the situation truly enviable and desirable. Young men out of employment or uncomfortable, "There is a tide in the affairs of men, which, taken at the flood, leads on to fortune"; nick in instantly and enlist.'

Many colonels regarded their regiments as personal property and were as often as not absentee landlords. They were content to draw the pay for the men shown on the muster rolls, many of whom were absent or in some instances dead, and leave it to some subordinate officer to lick the soldiers into shape. Discipline was brutal; in Boston in 1770 British soldiers were known as 'Bloodybacks' on account of the constant use of the lash. The average British officer was perfectly willing to die gallantly in battle at the head of his men but he had better things to do than waste his time on parades or drill when in a peacetime garrison. Not so Burgoyne, who had very firm views on the relationship of officers with their men, as he proceeded to set out in a Code of Instructions which was issued to his officers.

He began by comparing the systems of discipline which then prevailed in the Prussian and French armies. In the former men were trained like spaniels, 'by the stick'. The French on the other hand substituted honour instead of severity. 'The Germans are the best; the French, by the avowal of their own officers, the worst disciplined troops in Europe. I apprehend a just medium between the two extremes to be the surest means to bring English soldiers to perfection.' Burgoyne went on to show 'why an Englishman will not bear beating so well as the foreigners in question.' He emphasised the need to treat soldiers as 'thinking beings' and insisted on his officers 'getting insight into the character of each particular man.'

Swearing at soldiers was prohibited and some relaxation of the strict officer-man relationship was to be encouraged from time to time. 'There are occasions, such as during stable or fatigue duty, when officers may slacken the reins so far as to talk with soldiers; nay, even a joke may be used, not only without harm but to good purpose, for condescensions well applied are an encouragement to the well disposed, and at the same time a tacit reproof to others.' It may seem incredible to us today that officers should have needed to be encouraged to talk, and even to joke with their men, but this was radical stuff at the time it was written. What is equally surprising is that this relaxed attitude between officers and soldiers remained a characteristic of Burgoyne's regiment long after he was dead and forgotten.

On parade Burgoyne required proper subordination among officers, but off parade was another matter. There should be complete social equality in private intercourse, and 'any restraint upon conversation (off parade), unless when an offence against religion, morals, or good breeding is in question, is grating . . .' He went on to require his officers to study their profession, pointing out that 'A short space of time given to reading each day, if the books are well chosen and the subject properly digested, will furnish a great deal of instruction.' He urged his officers to learn French, because the best military books of that time were written in French, and to study mathematics because it was an essential subject for the military profession. 'An officer ought to write English with swiftness and accuracy,' and 'If a man has a taste for drawing, it will add a very pleasing and useful qualification; and I would recommend him to practise taking views from an eminence, and to measure distances with his eye. This would be a talent peculiarly adapted to the light dragoon service.'

Horsemanship was more or less taken for granted among officers who had ridden from their earliest days, but this was not good enough for Burgoyne. Horsemanship was to be studied, as was the fitting of saddles and bridles. 'I hope I shall not appear finical, if I recommend to officers sometimes to accoutre and bridle a horse themselves until they are thoroughly acquainted with the use of each strap and buckle.' Anticipating, no doubt, complaints from those who regarded such an instruction as beneath their dignity as officers, Burgoyne gently asks them to consider, 'whether a reproof from a field officer, or, what is perhaps worse, a criticism from a judicious spectator, would not give them more pain'? He even asks his officers to demean themselves so far as to acquire a knowledge of farriery, and to interest themselves in the feeding of their horses.

Burgoyne's regiment soon reached a high state of efficiency and was judged to be fit for active service. In April, 1761, two troops under Captains Sir William Williams and Sir George Osborne embarked for the expedition against Belle Isle, off the coast of Brittany. They seem to have distinguished themselves but Williams was killed. The following year Burgoyne and his regiment embarked for Portugal. The Portuguese, threatened by Spain and France, adhered to their old treaty with Britain, and an expeditionary force was sent to their aid and placed under the command of one of the leading mercenary generals of that time, the Count of Schaumberg Lippe. It was just the kind of campaign Burgoyne was looking for to blood his new regiment.

The 16th Light Dragoons sailed for Portugal in May, 1762, and remained there for a year. They were to find the army of their Portuguese ally a somewhat undependable instrument of war. At a banquet given in his honour by the Portuguese generals the Allied Commander-in-Chief was astonished to discover that the waiters were captains and lieutenants in the army he had come to command. Lippe held the Portuguese artillery in such contempt that he offered a prize for the gun team which succeeded in hitting the flag above his tent, and to make the competition more exciting he directed that it should be held during a return banquet he gave for the Portuguese generals. It must have seemed to Burgoyne and his regiment that there was little prospect of such a heterogeneous mob defeating the armies of France and Spain, but in fact the Spaniards were little better.

The 16th first distinguished themselves in an action at Valencia de Alcantara, just across the Tagus in Spain. Burgoyne, with a brigade composed of British and Portuguese infantry and his own regiment, surprised a superior Spanish force by a daring night march, culminating in the advance guard of the 16th Light Dragoons galloping into the main square, quickly followed by the rest of the regiment, and taking prisoner the Spanish general in command and most of the

Regiment of Seville, together with three stands of Colours. During the subsequent pursuit a detachment of six men of the 16th under a sergeant charged and took prisoner a Troop of 26 Spanish dragoons and Burgoyne became almost lyrical in describing this exploit in his despatch to Field-Marshal Lippe.

The exploits of the 16th Light Dragoons and their dashing Brigadier at Valencia de Alcantara were not, however, sufficient to prevent a Spanish invasion of Portugal. Shortly afterwards the Spaniards, in considerable strength, moved across the frontier but, we are told, 'the dauntless countenance of the British troops over-awed their opponents'! Over-awed or not, the Spanish army continued to advance and in consequence 'some retrograde movements were, however, necessary.' The 16th Light Dragoons covered the withdrawal with considerable *éclat* and eventually the Tagus was crossed. A force under Burgoyne took up a position at Villa Velha in October and again succeeded in surprising the Spaniards, which does not seem to have been all that difficult to achieve.

In this instance the Spanish had occupied an old Moorish castle on the north bank of the Tagus and had fortified two hills on the plain of Villa Velha. Burgoyne formed a small force of 16th Light Dragoons, Grenadiers and Portuguese and placed them under command of a certain Colonel Charles Lee who had won his spurs in North America fighting against the Indians and the French. Lee was a remarkable character who later threw in his lot with the American colonists and had several exchanges with Burgoyne, as well as an unfortunate experience at the hands of the 16th Light Dragoons. Although Lee began his service in the Infantry he became a dedicated cavalryman. After his service in Portugal he joined the Russian army and wrote to a friend – 'I am to have command of Cossacks and Wollacks, a kind of people I have a good opinion of. I am determined not to serve in the line [infantry]; *one might as well be a churchwarden.*' Good cavalry sentiments which have been echoed down the years!

At Villa Velha Burgoyne repeated his tactics at Valencia de Alcantara. Lee's force forded the Tagus, moved by forced march through the mountains, and fell on the unsuspecting Spaniards around two o'clock in the morning. The result could be predicted. Most of the Spaniards were shot or bayonetted in their tents and the few who attempted to make a stand were charged by the 16th Light Dragoons under Lieutenant Maitland and cut down almost to a man. With hardly any loss to themselves Lee's force captured guns, horses and men, and both he and Burgoyne figured prominently in the Commander-in-Chief's despatch. The war ended shortly afterwards and the 16th Light Dragoons and their Colonel returned to England with a considerable reputation. Lady Charlotte Burgoyne returned with them, for, like all good soldiers' wives, she had followed the drum to Portugal, despite her husband's protests.

On their return to England the 16th Light Dragoons were to find themselves very high in royal favour. The reason for this must have been largely due to John Burgoyne's popularity at Court, as well as to the success of the Light Dragoon experiment. King George III took his responsibilities as Commander-in-Chief of the Army extremely seriously and was never happier during the earlier years of his reign than when reviewing his regiments. On 20 May, 1766, he reviewed the 15th and 16th Light Dragoons on Wimbledon Common and was so pleased with what he saw that he commanded that the 15th should in future be styled 'The King's', and the 16th, 'The Queen's', Light Dragoons. Burgoyne's Light Horse therefore became Queen Charlotte's Light Dragoons and have ever since retained the title of 'The Queen's'.

Throughout history the subject of dress has had an obsessive interest for soldiers and this is as true today as it was in Roman times. It is hardly surprising therefore that the dress of his newly-formed Light Dragoons occupied a good deal of King George III's attention and by 1768 the 16th were dressed as follows. On their heads a helmet with a horse-hair crest. The coats were scarlet with blue facings. Waistcoats and breeches were white with black boots reaching to the knee, while the cloaks were scarlet with white linings and blue capes. There were other detailed instructions concerning saddle cloths and so on which need not concern us here, but Queen Charlotte's cipher was embroidered on several of the appointments

which was to have some significance 200 years later.*
Three Guidons were carried. These swallow-tailed flags consisted of the first, or King's, Guidon of crimson silk, and the second and third Guidons of blue silk. The two latter were embroidered with the Queen's cipher within the garter and bore the regimental motto, *Aut cursu, aut cominus armis,* which remains the same today. It can be translated as 'Either in speed, or in close combat', which summed up, reasonably accurately, the role of the Light Dragoon. The choice doubtless bears witness to John Burgoyne's classical education at Westminster School.

The light dragoon was lavishly supplied with weapons. The principal weapon was the sword, 36 inches long with a straight blade. Each man also carried two pistols with 9-inch barrels and a carbine 2 feet 5 inches long in the barrel; pistols and carbine were of the same calibre so that the same bullet fitted both. Neither carbine nor pistol was accurate for any great distance but when discharged point-blank could inflict a fatal wound. It was usual to dismount before using the carbine.

Parades and ceremonial have their part to play in the training of a soldier but they are, after all, only a means to an end, and in the soldier's case the end can only be fitness for war. On 18 and 19 April, 1775, the King's troops clashed with the North American Colonists at Lexington and Concord in Massachusetts and the first shots were fired in a war which was to last for eight years and end with the emergence of the United States of America as part of the community of nations. When it was finally decided to send reinforcements, the 16th Queen's Light Dragoons were among the regiments selected.

The 16th's destination was Boston, where their Colonel, Major-General John Burgoyne, had preceded them as one of the three Major-Generals sent out to assist General Gage. However, Boston

* When, in 1958, the Regiment's new Guidon was being designed, it fell to me as the Commanding Officer at that time to decide on the centre-piece. For various reasons it proved impossible to use the Regimental badge and motto and I finally suggested to Garter King-at-Arms that we should follow the same practice as in 1768 and have the cipher of Her Majesty Queen Elizabeth II, our Colonel-in-Chief, as the centre-piece, to which The Queen agreed.

was evacuated by the British on St Patrick's Day, 1776, and the convoy of ships carrying the 16th and other reinforcements was switched to New York. It had been a hideous voyage of over three months in the teeth of Atlantic gales. The ships were crammed with troops, horses and supplies. Below decks the stink was worse than in any stable and for much of the time the soldiers lived in almost total darkness. Food was short and almost uneatable. Drinking water soon went foul and deaths from dysentery and scurvy were frequent. In really rough seas the horses went mad with fear between-decks and often cast themselves. It then became necessary to fumble around in the darkness among the flailing hooves to release the cast animal; and if it was too badly injured to get back on its feet, to kill it then and there by cutting its throat. After a sea voyage of such length and under such conditions it is incredible that a regiment could be fit for anything, but within a week of landing in America the 16th Light Dragoons, commanded by Lieutenant-Colonel the Honourable William Harcourt* were in action at the battle of White Plains on 28 October, 1776.

Cavalry were in short supply in North America and in consequence the 16th Light Dragoons were given little rest. At this stage in the war the Colonists had had little chance to organize an army and were hard put to it to avoid a pitched battle with the British. Their commander-in-chief, General Washington, was a master of withdrawal tactics. He always managed to slip away just as the net was closing round him and this involved the cavalry in continual encounter actions. In November, 1776, the Colonists abandoned New York and fell back across New Jersey with the British in pursuit. Their objective was Philadelphia where the Colonists had established a government.

Probably the most enterprising American general at this time was that same Charles Lee who had led the 16th Light Dragoons at Villa Velha in Portugal only 14 years previously. Lee was a strange character who quarrelled with most people sooner or later. Moody, slovenly, and apparently incapable of affection except for the pack

---

* Later Field-Marshal Earl Harcourt and Colonel of the Regiment from 1779 to 1830.

of dogs which followed at his heels, Charles Lee was daring at one moment, and fumbling and slow the next. On 12 December, 1776, he put up for the night at an inn near Morristown in New Jersey. He had earlier boasted that he was 'going into the Jerseys for the salvation of America,' but a patrol of the 16th Light Dragoons fell in with a countryman who told them that Lee was nearby. At 10 o'clock the next morning, having written despatches and breakfasted, Lee was about to come down to mount his horse when the inn was surrounded by about 50 men from the 16th. They were commanded by Lieutenant-Colonel Harcourt, who had served under Lee in Portugal, and it must have been a strange reunion. Harcourt wrote that he had captured the 'most active and enterprising of the enemy generals' and at one time it was thought that Lee's capture might result in wholesale American defections. But it did not happen that way. Lee was at first in some danger of being hanged as a rebel but was later exchanged for a captured British general.

For the next three years the 16th Light Dragoons marched and counter-marched through the Jerseys and Pennsylvania. They were present when the British entered Philadelphia and took part in the battles at Brandywine Creek and Germantown. They scoured the countryside for supplies and annoyed both loyalists and rebels alike by grazing their horses in the cornfields. It was hard campaigning since it was often difficult to distinguish friend from foe. There were, in fact, too few troops to hold down such vast areas of country, much of it almost primaeval wilderness. In 1777 the British conceived a plan to send an army down from Canada to join forces at Albany on the River Hudson with the troops already operating out of New York under Generals Howe and Clinton. The command of this army was entrusted to General Burgoyne. It was a well-found little army of British and German troops, but war in the North American wilderness was totally different from campaigning on the plains of Germany. Burgoyne was a good general, probably the most imaginative British general in America, but he eventually had to surrender his troops to superior American forces at Saratoga on 17 October, 1777. Almost immediately afterwards France entered the war on the side of the Americans, as did Spain, and Britain did not possess the resources to deal with all her enemies. Before the end came the 16th Light Dragoons, by now virtually horseless, were sent home in 1779 where they learnt that General Burgoyne had resigned the Colonelcy of the Regiment in protest at his treatment by the British government after Saratoga. 'Burgoyne's Light Horse' would have to find another nickname.

Chapter 3
Peninsula

IN 1789 the fire which had long been smouldering in France broke into flames and soon the rest of Europe became uneasily aware that the most civilized and powerful of all the European nations was in the grip of revolution. The British had little cause to love the French, and least of all the Bourbons, but if Louis XVI's head was to roll today, who could say when it might not be George III's? First Austria and Prussia, and then more tardily Britain, joined hands to support the royalist cause, and in the Spring of 1793 an expeditionary force of 10,000 men sailed from England for the Continent. The 16th Light Dragoons, their strength augmented by raw recruits who scarcely knew how to ride, formed part of this force under the command of the King's second son, the Duke of York. It marked the beginning of a war which was not to end until Napoleon was defeated at Waterloo.

In 1793 the 16th Light Dragoons were commanded by Lieutenant-Colonel Sir Robert Laurie, a baronet who had paid around £5,000 at regulation price for the privilege of commanding the Regiment. They marched from Ostend to Tournai where an Austrian army was already assembled and there followed a long and inconclusive campaign against the armies of Revolutionary France. Undisciplined though the French may have been, they were inspired by a determination to spread the revolutionary gospel. They even fought at night and did not stop fighting when winter came and other European armies went into winter quarters until the spring. It was during this fighting along what is now the frontier between France and Belgium that the 16th Light Dragoons won their first Battle Honour – at *Beaumont* (near Le Cateau) on 26 April, 1794. Although the Duke of York commended the conduct of the 16th Light Dragoons and a regiment of German Hussars as being 'beyond all praise', it was not an action of much importance, but it was to be the first of many Battle Honours.

These skirmishes along the frontiers of France were inconclusive affairs but the Duke of York was compelled to retreat to the safety of the walls of Tournai where on 10 May, 1794, he was attacked by the French. More by luck than good management – there was an almost total absence of good management in that campaign – the French were beaten off and the 16th Light Dragoons distinguished themselves by charging a battery and taking the guns. This spirited little action won them another Battle Honour – *Willems*. From then on the campaign went from bad to worse and is chiefly memorable for the doggerel verse about 'The Noble Duke of York,' and the fact that it was during this succession of blunders that the young Arthur Wellesley (later the Duke of Wellington) won his spurs while commanding the 33rd Foot. When, many years later, he was asked whether he had found the campaign useful, he replied: 'Why – I learnt what one ought not to do, and that is always something.'

The unconventional tactics and the revolutionary ardour of the French got the Allies on the run and in the case of the British they began a retreat which did not end until they reached Bremen. The winter was the most severe in living memory, and the Duke of York went home fairly early in the proceedings, as did some other senior officers. This not very glorious episode in the Regiment's history ended with their embarkation for England in February, 1796.

The Regiment was commanded in 1803 by Lieutenant-Colonel Stapleton Cotton who was later Wellington's principal cavalry commander in Spain. Later still, Cotton, by then Viscount Combermere, was Commander-in-Chief in India when the Regiment distinguished themselves under his command at Bhurtpore. Cotton was popular but he did not possess much common sense or ability. Many years later the Duke of Wellington, by then an elder statesman, was asked for advice on a commander to lead an expedition against Burma and he suggested Combermere. 'But, Your Grace,' expostulated the questioner, 'I always thought you considered Lord Combermere a fool?' 'So he is,' replied the Duke, 'and a damned fool, but he can take Rangoon!'

While the 16th Light Dragoons were engaged in putting down rebellion in Ireland, hunting smugglers in Kent, and taking part in royal reviews, Napoleon was extending his control over Continental Europe. The Portuguese asked the British government to come to their aid. After the usual chapter of disasters, which marks the beginning of almost every British campaign, the British began to build up an army based on Lisbon and commanded by the 40-year-old Sir Arthur Wellesley. The 16th Light Dragoons joined this army on 15 April, 1809, and did not return home for five years, in the course of which they were present at seven pitched battles, losing 300 officers, NCOs and men, and 1,416 horses.

They began the campaign under the command of Colonel the Hon George Anson. One of the subalterns was William Tomkinson, son of a Cheshire squire, whose diary of those eventful years is probably the best account of life in a cavalry regiment during the Peninsular War. Tomkinson was described by a brother officer as 'one of the best soldiers I have ever met,' and although he was a countryman at heart, always longing to get back to his estate at Tarporley, he was also a keen student of his profession. In this he was unusual since most British cavalry officers believed their principal task in war was to charge at the enemy and die sword in hand.

Wellington's campaign in Portugal and Spain lasted for five years with fluctuating fortunes. By the end of the campaign the regiments involved in it had reached a peak of efficiency unequalled in the British Army since Marlborough's time. The 16th Light Dragoons' first action was at Albergueria Nova on 9 May, 1809, and from then onwards they were almost continuously employed in outpost duties. They took part in the battles of *Talavera, Fuentes d'Onor, Salamanca, Vittoria*, and *Nive*, all of which are emblazoned on their Guidon, as is *Peninsula*. They covered Wellington's withdrawal to the Lines of Torres Vedras in the autumn of 1810 and followed up the retreating French army in the spring of the following year. 'We soon learnt to sleep in the day or night,' wrote Tomkinson. 'We never undressed and at night all the horses were bridled up, the men sleeping at their heads, and the officers of each troop close to their horses.'

It was a high-spirited army, in which the cavalry set the pace for gaiety. When they were not fighting there were fox hunts and balls, as well as much wining and dining and courting of the senoritas. For weeks on end the cavalry slept out in all weathers on the outpost line, lucky if they could find temporary shelter from the snow or rain in a peasant's flea-ridden hut. The officers had to be tough, for the men they commanded were astonishingly so. One soldier had a leg and an arm amputated without any kind of anaesthetic and was found the following morning propped up on his remaining elbow calmly smoking his pipe. The women who followed their husbands were equally tough. One of them, Bridget Skiddy, carried her man, knapsack, musket, and all, when he could march no farther during a retreat.

The Hon Edward Somers-Cocks, who commanded a Troop of the 16th, made a great name for himself until he was killed in the breach at Burgos in 1812, by which time he had transferred to the 79th Highlanders. He was even mentioned in a dispatch in 1810, a rare honour since Wellington was sparing in his praise. Tomkinson wrote of him, 'the men in his troop were very fond of him, and would hollo when in a charge, "Follow the Captain! Stick close to the Captain." They called his squadron, "The Fighting Squadron".' Another regimental character was Sergeant-Major Blood, who behaved so gallantly at Tudela that Wellington offered him an immediate commission. Blood refused it and Wellington gave him 100 dollars instead. After the French defeat at Vittoria, Blood and six of his men managed to loot 6,000 silver dollars. He later obtained a commission as a riding master, dying in 1840, aged 64. There is a memorial to him in Cheadle Church which was subscribed to, among others, by Lord Combermere and the Duke of York, the last stating, 'H.R.H. thinks it right to state that Mr. Blood is one of the most meritorious old officers in the King's service.'

Cavalry officers in Spain used to assert that the main purpose of cavalry was to give tone to what otherwise would be simply a vulgar brawl. This did not endear them to the rest of the army. 'Perhaps I need not tell the reader,' wrote Gleig, 'that between the infantry and cavalry in the British Army a considerable degree of jealousy exists;

the former description of force regarding the latter as little better than useless, the latter regarding the former as extremely vulgar and ungenteel.' Wellington used to complain that his cavalry 'could gallop, but could not preserve their order,' but he had no cause to complain of their handling of outpost duties. It has been written of the 16th Light Dragoons' record in the Peninsula:

'When the Regiment formed the picquets the army behind it slept secure, for in no single instance were its outposts surprised; when it was on reconnaissance no General need hesitate for want of information; when employed to harass the enemy's outposts no French picquet rested in peace. In advance or in retreat, in quarters or in the field, the conduct and discipline of the 16th during the Peninsular War was all that could be desired and the Regiment gained honour and distinction for every officer who was fortunate enough to command it.'*

---

* *History of the 16th Light Dragoons (Lancers)* by Henry Graham (privately printed), 1912.

Chapter 4
Waterloo

AT the end of 1814 the 16th Light Dragoons returned from France to Hounslow. Soon afterwards they found themselves involved in skirmishes with the London mob rioting in protest against the passage through Parliament of the Corn Laws. The dragoons were heartily booed as they rode through the streets on their way to Westminster. It must have come as a relief from such distasteful duties when news was received of Napoleon's escape from Elba. As so often in British history there was a hurried attempt to raise an army, and on 11 April, 1815, the regiment embarked at Dover and Ramsgate. The transports were small colliers ill-suited for the carriage of horses but fortunately the sea was calm. On arrival at Ostend the animals were slung overboard and left to their own devices to swim ashore. They were escorted by troopers who landed naked in full view of the scandalized Belgians.

The 16th soon discovered that campaigning in the Low Countries was a distinct improvement on Spain and Portugal. 'We spent our time visiting the old churches and museums,' recorded Lieutenant John Luard in his diary, 'and there were many balls given in our honour'. According to Tomkinson the soldiers were similarly fêted. 'The men cannot stand the good treatment they receive from the persons on whom they are billeted,' he wrote, 'and some instances of drunkenness have occurred.' Lieutenant-Colonel James Hay, the 16th's commanding officer, took his officers to visit the battlefield of Oudenarde. He asked Luard for his comments on the battle, which was not one 'I knew anything about,' said Luard. 'My answers were at best guesses. Since the colonel did not question them I can only conclude he knew little more about the battle than I did.'

During April and May the allied armies concentrated in Belgium under the Duke of Wellington. Little news came out of France. 'We heard little information,' wrote Tomkinson, 'and the present distribution of the troops appears more like a distribution for winter quarters than for an approaching campaign.' Major-General Sir John Vandeleur, who commanded the cavalry brigade consisting of the 11th, 12th and 16th Light Dragoons, had arranged a field day for 16 June. The 16th's adjutant was sick and Hay appointed Luard to act in his place.

He was wakened at 5 am on 16 June. The colonel wanted to see him immediately. Mounting his horse he set off at a fast canter to Denderwinche where regimental headquarters had been established in an inn. On the way he passed Tomkinson riding hard in the direction from whence Luard had just come. He told Luard that he had fallen in with a Belgian dragoon in Enghien carrying dispatches from the frontier. The French were on the move towards Charleroi and the Prussians were advancing to meet them. Hay had already heard the news and on Luard's arrival ordered him to assemble the regiment. It was midday before the 16th Light Dragoons received their orders. They were to march to the crossroads at Quatre-Bras. By 2 pm they had reached Nivelles where they met the first wounded and stragglers from the battle. Now they could hear the booming of the guns, and the sound caused Hay to order the regiment to trot. Then Hay, taking Luard with him and following after Vandeleur and his staff, went cantering ahead towards Quatre-Bras. But the light was fading by the time the 16th reached the battlefield; there was little to see except the occasional flash of musket or cannon, and the winking fires of the outpost line. The night air stank of gunpowder and the ground was littered with the breastplates of the French Cuirassiers. Time and again they had charged gallantly, and been thrown back equally gallantly, by the rocklike squares of the 42nd, 79th and 92nd Highlanders.

John Luard slept that night in a cabbage patch a few miles north of the cross-roads. Around him lay the weary troopers wrapped in their cloaks, stiff and sore after twelve hours in the saddle. Their horses were equally weary, hardly able to summon up the strength to nibble at their hay. Both horses and men were thirsty. There were no streams and the few wells had long since been drained dry. The next morning dawned dull and lowering. Luard rode down the road towards the outposts. There was no sign of movement from the French and the smoke of their bivouac fires showed they were busy preparing *petit dejeuner*. The enemy had still made no move by ten o'clock and Wellington himself rode round the outposts to see things for himself. But meanwhile the main body of the British Army, protected by a screen of cavalry outposts, was withdrawing to a previously selected position at Waterloo, twelve miles north of Quatre-Bras and covering Brussels.

It was not until the early afternoon that Napoleon discovered what was happening. He was furious with Ney for failing to press hard on the heels of the retreating British. After first sending d'Erlon with his cavalry corps to break up the British withdrawal, he eventually galloped forward himself. He was too late. At 3 pm it began to rain in torrents, turning the ground into a bog and quenching the fires of the French pursuit. 'It became impossible for the French cavalry to press our columns in any force,' wrote one of the English rearguard. 'In fact, out of the road in the track of our own cavalry the ground was poached into a complete puddle.' Gradually the pursuit tailed off, and as night fell the drenched and saddle-sore cavalry passed through the equally wet infantry, drawn up on the high ground to the south of the village of Waterloo. The duties of adjutant kept Luard busy until well after midnight. While on his rounds he met a soldier coming out of a cottage with a grandfather clock on his back. He enquired what the man intended to do with it. 'If you come to our troop,' he replied, 'you'll soon see what I will do with it. I'll make the beggar tick.' As Luard passed that way later he was invited to warm himself before a blazing fire of which the clock formed the principal ingredient. When at last Luard was able to lie down in his cloak, he found it difficult to sleep: 'My breeches were soaked and the discomfort kept me awake for most of the night, although I was extremely fatigued.'

For most of the night it rained fitfully. Then, as the grey sky slowly cleared, a watery sun came out, and Sunday, 18 June, began. Unshaven, bedraggled, and covered in mud, the 16th Light Dragoons struggled to their feet and went off to feed their horses. It had been a night to remember. 'At nine o'clock I went to Colonel Hay,' wrote Luard, 'and he ordered me to mount the regiment, which I did. My horse was so cold and shivered so much that he could hardly stand for me to mount. The regiment was placed on the left of the position, and we dismounted under a rising ground, waiting for orders.'

They could see the French clearly. Teams of horses were struggling forward through the mud to position the artillery. The sight proved too much for a veteran named Price, the shoemaker in Tomkinson's Troop. Dismounting hurriedly, he moved off to the rear. No-one tried to stop him and he did not rejoin until after the battle. He was a regimental character and his comrades bore him no malice. At eleven o'clock the 16th were moved to a position to the east of the Brussels highway. The rising ground to their front was occupied by infantry and very little happened there for the next hour or so. The French cannonade began at 11.35 am but the ground was so sodden that many of the shells buried themselves where they landed.

Waterloo was fought within an astonishingly small area. From north to south the battlefield extended for less than four miles, and it was rather less in width. Within this restricted space there were deployed more than four hundred pieces of artillery and over 140,000 men. Napoleon launched attack after attack on the British lines drawn up on the ridge. Each assault was thrown back. The 16th had mounted and moved off to break up the withdrawal of Durotte's Division after one of these attacks. They were recalled to charge with the rest of Vandeleur's cavalry brigade in an attempt to rescue Ponsonby's brigade of heavy cavalry. Ponsonby's – the Royals, Greys and Inniskillings – had got out of control after charging a retreating French column. As they rallied after the charge they were attacked by French lancers intent on their annihilation. Colonel Hay's trumpeter sounded the *Charge* and the 16th Light Dragoons plunged forward down the slope into the smoke-filled valley below. Luard felt his horse stumble, pulled it up again by main force, and at the same moment saw his colonel pitch forward out of the saddle. Swerving to avoid him, Luard passed on to cut down a French lancer looming out of the smoke. He swung his horse to avoid a bayonet thrust and the battlefield dissolved into a series of individual combats as English light dragoon fought French lancer, isolated from their comrades by the smoke, the noise, and the general confusion. The French broke away, the 16th rallied to the shrilling of their trumpeters and withdrew with the rest of Vandeleur's brigade to their former position.

The 16th brought back with them their wounded colonel. He had been shot from behind, probably by a stray shot from his own infantry, and the wound seemed mortal. But Hay recovered and lived to become a general. The 16th had suffered few other casualties. The smoke was now so dense that men only knew what was happening in their immediate vicinity. At about 4 pm they were told that the Prussian army was arriving on the battlefield. At about the same time a bullet hit Luard's charger in the head and it collapsed under him. He transferred to a trooper's horse as the 16th moved to the west of the Brussels highway, passing the 18th Hussars in the smoke. 'I saw my brother George,' wrote Luard, 'the 18th Hussars being close to us. While in this position I was talking to Lieutenant Philips of the 11th Light Dragoons when his head was shot off by a cannon shot . . . The Belgians began to give way, the enemy's fire being too hot for them, and we closed our squadrons and would not let them go to the rear. Sir John Vandeleur and I moved to the front and encouraged them. The fire then slackened and they held their ground.' It was a critical moment and as so often the Duke himself turned up at the right time and in the right place. 'That's right, that's right,' he called out to Luard, 'Keep them up! Keep them up!' And placing two fingers to his cocked hat, the Duke of Wellington, 'his face blackened with smoke, but otherwise his appearance as neat as ever,' cantered off to shore up some other section of his battered line.

Napoleon was about to try a gambler's throw. The flower of the French army, the Imperial Guard, advanced to wrest victory from defeat. At their head was a hatless Marshal of France, Ney, as smoke-blackened as Wellington, and above them waved the Colours emblazoned with Marengo, Austerlitz, and all the other victories. They were met by the musketry of the 1st Foot Guards and the 52nd Light Infantry which broke their ranks. *Le Garde recule!* The

French infantry, watching horrified as the Imperial Guard fell back in retreat, began to retire. As they retreated across the fields littered with the debris of battle, Wellington gave the signal for a general advance.

The 16th Light Dragoons trotted forward. The trumpets sounded. Vandeleur's brigade poured down the ridge. At first there was no-one to sabre, for the French were retreating too quickly, but soon the regiment came up against formed bodies. 'The enemy's infantry behind the hedge gave us a volley, and being close to them, and the hedge nothing more than some scattered bushes without a ditch, we made a rush and went into their columns . . . they running away to the square for shelter.' Tomkinson led his Troop in several such skirmishes as the French withdrew and in one of them Captain Buchanan fell, as did Lieutenant Hay in another. As night fell, the Prussians took up the pursuit and the weary British withdrew to count the cost. 'Next to a battle lost, the greatest misery is a battle won,' wrote Wellington the next day to Lady Shelley, thinking of all his friends who would never return from Waterloo. And so thought many a 16th Light Dragoon as he listened to the roll call that June night. But it was a glorious victory.

Chapter 5
The Scarlet Lancers

THE Regiment returned from Waterloo to Romford in Essex, but in January, 1816 they were sent to Ireland where according to John Luard, 'Several of the officers fell in love . . . Colonel Pelly married Miss French – Major Persse married a Miss Moore – it was a very hospitable quarter and the officers of the 16th were paid a great attention.'*

While in Ireland the 16th were armed with the lance after the model of the Polish Lancers in Napoleon's service, and shortly afterwards their title was changed to the 16th, The Queen's, Light Dragoons (Lancers) which soon became shortened to the 16th The Queen's Lancers. Both George IV and William IV were constantly altering Dress Regulations and in the cavalry uniforms became more and more gorgeous and less and less practicable – 'Frenchified,' as the Duke of Wellington contemptuously said. Lancer regiments copied the dress of the Polish Lancers, including the head-dress with drooping red and white cock-tail. At first the uniform was blue, but in 1830 it was changed to scarlet. There was yet another change in 1846 when all Lancer regiments, except the 16th, were ordered to wear blue. The 16th consequently become known throughout the cavalry as the 'Scarlet Lancers.'

While quartered in Sheffield in 1821 the Regiment came under the displeasure of the Court. At the time George IV was involved in an unseemly wrangle with his consort, Caroline, and was scheming to prevent her being present at his coronation. The officers of the 16th, a Queen's regiment, made no secret of their sympathies. They invited

---

* There was a similar increase in the marriage stakes in 1958 when the Regiment arrived in Catterick after 10 years overseas.

the gentry of the West Riding to dine with them, and loudly, and perhaps with unnecessary frequency, toasted the 'Queen'. As might have been expected, reports soon reached the ears of the King, and in 1822 the Regiment received orders for India. It would be 24 years before it returned home, and only one officer who embarked with the Regiment at Gravesend in 1822 was still serving with it in 1846. He was George Macdowell who went to India as a junior captain and returned as a lieutenant-colonel and commanding officer.

The Regiment's destination was Cawnpore which they reached by river-boat up the Ganges. These boats were hauled upstream by their crews, pulling and pushing the unwieldy craft over the sandbanks. Each night they tied up to the bank and the soldiers stretched their legs in the cool of the evening. Ten miles a day was about the average. It took the 16th Lancers nearly four months to reach Cawnpore from Calcutta. No concession was made to the climate so far as dress was concerned and the high-necked scarlet tunics were black with sweat. As they travelled farther and farther into the heart of India, the country grew flatter and more dusty. 'Hotter than Hades, and a damned sight less interesting,' is how one troop sergeant described it.

They were not all that much better off when they reached Cawnpore. Officers could find plenty to do during off-duty hours but the soldier was not so fortunate. 'In the first place,' writes Sergeant Thomas, 'a troop is huddled together in one barrack to the strength of 87 men. These have continual intercourse with one another as they are ranged at each side of the barrack . . . The duty of a dragoon is very easy on account of the great number of followers. Each man has a syce to clean and saddle his horse. Each troop has two barbers, two shoe blacks, two belt cleaners, and eight dhobies or washerwomen . . . Thus the men have nothing to do except field service during the cold season. The cold months glide rapidly past and the European is recruited to health and vigour, but he must prepare early in the month of April for his term of imprisonment – the hot winds, suffocating dust and doors closed with tatties, the barrack room being watered from outside by natives to keep them cool. At this time the men know not how to pass away the time unless by drinking or gambling; thus they are led to be drunkards or gamblers before they have been many years in India.'

Sergeant Thomas blamed much of this moral collapse on the regimental canteens which 'deprive the soldier of his comforts while realising a fortune in two or three years for the canteen sergeant.' He mentions one such sergeant who made £1,400 in the space of 18 months. The mortality in India from the climate, disease, drink and vice was appalling, and yet soldiers were hardly worse off than they would have been at home. Many of them volunteered to remain in India when the time came for their regiment to leave for home. Their pay was only a shilling a day but it went further than in England, and there was in addition an allowance of two drams of rum a day. When off duty soldiers were permitted a considerable degree of freedom and many of them chose Indian women as partners. Their own countrywomen were in short supply and in great demand. A married man in the 13th Light Infantry died from cholera. His wife attended the funeral, and on leaving the cemetery was proposed to by another man in the regiment, to whom she had to reply that she was *already* engaged.

It is hardly surprising that officers and soldiers alike should welcome active service as a break in their monotonous existence. The 16th had their opportunity in 1825. The Rajah of Bhurtpore had seized the throne from his infant-nephew and defied the East India Company which had recognised the child's right to the throne. Bhurtpore, known as the 'Bulwark of Hindustan', was one of the strongest fortresses in India and was the capital of the Jats, Hindu yeomen who make excellent soldiers. They had successfully thrown back Lord Lake in 1805 when he made five successive attempts to storm the city. The British were determined to avenge this defeat. Lord Combermere, commander-in-chief of the Bengal Army, took command in person, and a considerable train of artillery was collected. The 16th received their orders to march on 5 November. It had been a very bad hot weather in Cawnpore, five men committing suicide. When the regiment was marching out of cantonments to join the Bhurtpore Field Force a dragoon drew his pistol and blew his brains out.

The force reached Agra on 3 December where Combermere joined it. He immediately reviewed the cavalry and the 16th were delighted to see their old commander again. They turned out at midnight to give him three cheers. He may have been uncommonly stupid but he had a good way with soldiers. The 16th cheered him again on 8 December when he gave orders for the advance to Bhurtpore. The balls protecting the steel tips of the lances were removed and the 16th led the advance, arriving under the forbidding walls of Bhurtpore two days later. 'On the 10th,' wrote Luard, 'Colonel Murray, who commanded our Brigade with four guns of Horse Artillery turned out at half-past 3 a.m. The infantry remained in camp. We proceeded to Sesma, then brought our left shoulders up and led straight for Bhurtpore . . . I was ordered to command all the skirmishers . . . I was ordered by Colonel Murray to cut off any enemy I could. I led the skirmishers close under the walls, while Skinner's Horse under the command of Mr Fraser, a civilian, made a sweep to the right. Some of the enemy's horse encamped under the walls retired as we advanced, but another party encamped further out were attacked by Fraser and driven towards one of the gates of the fortress, while I galloped on with my skirmishers and intercepted them as they approached the gate of the fort. We killed and wounded about 50 and took 100 horses . . . The guns of the fort now opened up . . . The skirmishers were then called in. Had I been supported by Infantry, I could have galloped into the fort with the retiring enemy horse.'

There are two interesting points about the 16th's encounter outside Bhurtpore. It was the first time in the history of the British cavalry that the lance was used in battle. The other is that during the night march to Bhurtpore a dragoon fell down a well in the dark, horse and man together, but they were hauled up without damage. Combermere now laid siege to the fortress and the 16th had little to do. The Jats kept up a steady fire from the walls and this fire became noticeably more accurate after Christmas Day when a sergeant of the Bengal Artillery named Herbert deserted to the enemy. Finally Bhurtpore was stormed on the morning of 18 January, 1826, after a mine had blown a breach in the walls. Combermere had to be forcibly restrained from accompanying the stormers. The fortress was not surrendered until four in the afternoon and only after fearful carnage. The Jats, wrote Luard, 'fought individually to the last, yielding their guns only with their lives.' He also paid tribute to the gallant conduct of the Company's Sepoy Regiments. The Rajah, Darjan Sal, was captured when trying to escape, and the deserter, Herbert, was hanged from the highest battlement. Lieutenant Mackinnon of the 16th witnessed the execution and wrote in his diary: 'The numerous spectators present can bear witness to the prolonged sufferings of the culprit. The rope being adjusted, one native pushed him off a low cart under the gibbet, while two others tugged at the rope to hoist him up. The convulsive writhings of the sufferer long haunted me. They lasted nearly twenty minutes.'

The Bhurtpore Field Force was broken up a few days later and the 16th Lancers marched to Meerut. They had lost no-one killed at Bhurtpore and they more than replaced their killed horses by those taken from the Jats. There was also prize money. Lord Combermere's share was £60,000, and each lieutenant-colonel received £1,500. Luard received £450 as a captain, while subalterns received £250. European sergeants were given £12 and the rank and file £4 apiece. The officers subscribed £1,000 for each of the widows of the four European officers killed in the battle; they also subscribed £1,000 to be distributed among the widows and orphans of the 61 European soldiers killed at Bhurtpore. By the standards of the time it was not ungenerous.

The 16th were still at Meerut in 1838 when they were ordered to join the 'Army of the Indus' at Ferozepore for the invasion of Afghanistan. The British, alarmed by the prospect of Russian influence becoming predominant in Afghanistan, decided to send an army to Kabul to overthrow the Amir, Dost Mahommed, and restore to the throne Shah Shujah, who had been ousted by Dost Mahommed. It turned out to be a disastrous campaign. Dost Mahommed was much more popular than Shah Shujah, and although the expeditionary force reached Kabul without much difficulty, Shah Shujah was incapable of governing his unruly people. Eventually the British garrison was compelled to retreat and was virtually annihilated in the passes between Kabul and India.

A great number of non-combatants accompanied the army to Afghanistan. There were men to pitch the soldiers' tents, since no European in India could be expected to pitch his own; grooms to tend the soldiers' horses, and grass cutters to cut grass for them; drivers to twist the tails of the patient bullocks which dragged the carts piled high with camp furniture, boxes of mess plate, crates of crockery, boxes of wine and cigars, and every other luxury required to make the campaign tolerable; men to lead the long strings of camels carrying hay for the horses; other men to drive the officers' buggies, to cook the food, to cut hair, to draw water, to clean the camp sites, to sell the soldiers food, drink and women. The 16th had nearly 5,000 followers to administer to their needs, and they were only one of many regiments. They also took a pack of fox-hounds with them to fill in time when they were not fighting.

There was in fact very little fighting. Kandahar surrendered without a battle. Ghazni, one of the strongest fortresses in Afghanistan, was stormed on 23 July, 1839, and Dost Mahommed fled to the mountains from Kabul. At 4 pm on 7 August, Shah Shujah, escorted by a squadron of the 16th Lancers, made a state entry into Kabul where he was ignored by his subjects. Nevertheless, the 'Army of the Indus' had not done too badly. It had marched 1,500 miles across desert and barren mountains; it had defeated the Afghans whenever they had made a stand; and it had restored Shah Shujah to the throne. The time had now come to enjoy themselves and one of the first acts was to lay out a race-course and a cricket pitch. The climate was good, the countryside pleasant, and the local ladies were handsome and in some cases forthcoming. On the surface everything seemed satisfactory, but the fires were smouldering beneath.

The amenities of Kabul soon palled. Officers and men who wandered too far from the camp were wounded or murdered. The 16th Lancers buried their commanding officer in Kabul. Lieutenant-Colonel Robert Arnold had been ill ever since leaving Kandahar and was carried in a litter to Kabul where he died. His effects were auctioned a few days later, sherry and bottles of sauce selling for the equivalent of £5 a bottle, while mustard was sold for £7 a square bottle. Arnold, who was very popular, had never properly recovered from a wound received at Waterloo. The 16th lost two other officers in the campaign, as well as 83 soldiers and 233 horses, mostly from disease and the privations due to a long march through barren country.

On 8 October, 1839, a General Order was published breaking up the 'Army of the Indus'. The 16th Lancers, among other regiments, were ordered to return to India, which they did via the Khyber Pass. The march took nearly four months and John Luard, by then serving as a staff officer in Calcutta, records with satisfaction, 'the safe return of the 16th Lancers to Meerut, and with them some foxhounds they had taken with them when they set out from that place.' The only serious incident occurred after the Regiment had arrived in India. When they arrived at the River Jhelum, about 400 yards wide, some 30 boats had been collected to transport the baggage, and the soldiers too, if required. But staff officers on the spot said the river was fordable, although the current was swift and the stakes marking the ford were not visible. Captain Hilton crossed the river and returned to say the river was fordable. The Brigadier then ordered the regiment to cross and they entered the water in column of threes. The advance guard reached the opposite bank safely but the main body lost direction and were swept away by the current. Confusion was made worse by baggage camels which entered the river higher up and were swept away. Captain Hilton was drowned, and with him 10 soldiers and 12 horses. It was an unnecessary disaster and there was much indignation with the staff for allowing the crossing to take place.

The 16th Lancers arrived in Meerut on 18 February, 1840, having marched 2,483 miles in 463 days, of which 212 days had been spent on the march, the balance having been halts in Kabul and other places. The fox-hounds must have been very foot-sore! Meanwhile in Afghanistan things went from bad to worse. Although Dost Mahommed surrendered to the British and was sent to Calcutta in exile, his supporters in Afghanistan kept the country in a state of rebellion. The British in Kabul seemed unaware of the ticking bomb under their feet and matters were not improved when Major-General William Elphinstone arrived in the autumn of 1840 to command the Kabul garrison. Elphinstone was a martyr to gout and knew nothing

1 *Her Majesty Queen Elizabeth II, Colonel-in-Chief of the Regiment, arriving at the bicentenary dinner and ball in 1959, accompanied by the late Brigadier P. E. Bowden-Smith, Colonel of the Regiment, 1944–59.*

2 General John Burgoyne, 16th Light Dragoons.

3 An officer of the 16th Light Dragoons by Thomas Gainsborough.

4 A private soldier of the 16th Lancers on patrol in the Peninsula.

5 (Left) Lieutenant-Colonel Rowland Smyth, CB, who led the charge at Aliwal.

6 (Below) The Battle of Aliwal, 28 January, 1846. Painting by Orlando Norie.

7 The 16th Lancers at exercise on Laffan's Plain, Aldershot, 1888.

8 5th Royal Irish Lancers; The Regiments' Mounts in 1880.

9 King Alfonso XIII of Spain. Colonel-in-Chief of the Regiment, 1905-1939.

10 Field-Marshal Viscount Allenby.

11 General Sir Hubert Gough.

12 Re-entry into Mons by the 5th Lancers, 1918. Painting by R. Caton Woodville.

13 *King George VI inspecting the Regiment prior to embarkation for North Africa, 1942.*

14 *Valentine Tanks moving into battle in Tunisia.*

15 Sherman Tanks moving forward at Cassino.

16 The advance to Castiglione.

17 *The Guidon presented to the Regiment in 1959.*

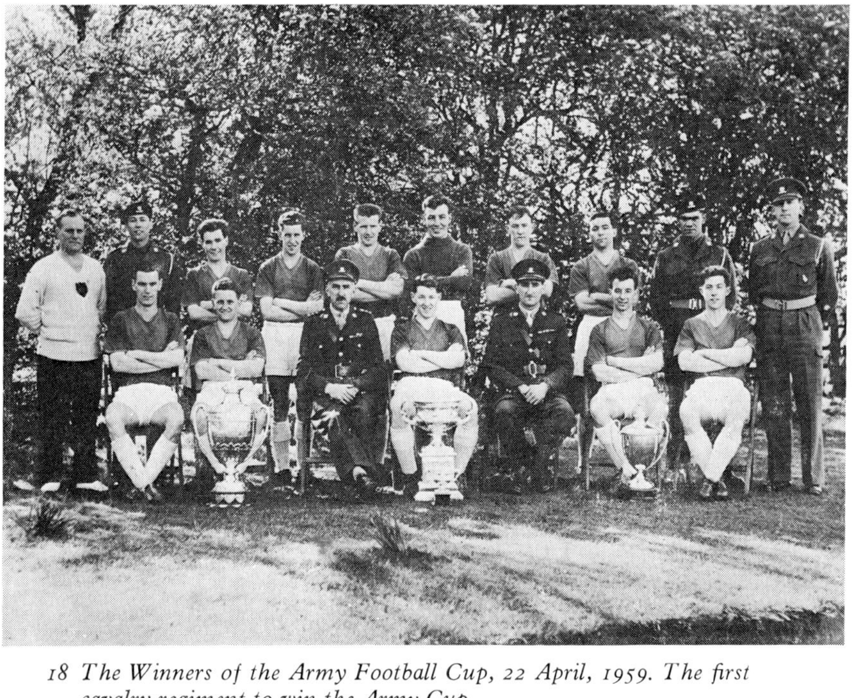

18 *The Winners of the Army Football Cup, 22 April, 1959. The first cavalry regiment to win the Army Cup.*

19 The Regimental Mounts in 1972.

20 Ulster, 1690: *The Battle of the Boyne.*

21 Ulster, 1972.

22 & 23 Border Patrol in N. Ireland, 1972. The explosion and the result.

of the East. He was well-connected and began life in the Guards, commanded the 33rd Foot at Waterloo where he was wounded, and commanded the 16th Lancers in 1822. John Luard describes him as 'a very gentlemanly and agreeable person, but from want of decision and never trusting to his own opinion unfit to command even a regiment.' When the Kabul mob rose against the British during the last days of 1841, Elphinstone proved to be incapable of dealing with the situation. Eventually the Kabul garrison of 4,500 soldiers and 12,000 Indian followers retreated to India under a safe-conduct from the Afghan chiefs which was not honoured. The column was massacred in the passes and only one man, Dr. Brydon, managed to reach safety. Elphinstone surrendered to the Afghans and died in captivity. It was the worst disaster suffered by the British in the East until Singapore one hundred years later.

The 16th Lancers were fortunate in having escaped the disaster. Two years later, in 1843, they took part in a minor war against the great Mahratta chieftain, Scindiah of Gwalior. It was not an arduous campaign, but they lost two men killed and seven men wounded, chiefly from artillery fire, in a battle at *Maharajpore* which was subsequently granted them as a Battle Honour. The bronze stars issued for this campaign were made from the metal of captured Mahratta cannon.

Chapter 6
Aliwal

MOST regiments choose to celebrate one out of their many Battle Honours. In the Regiment's case it is the famous charge at Aliwal during the First Sikh War. Aliwal itself is an obscure Punjabi hamlet of mud-brick houses a few hundred yards from the south bank of the River Sutlej in India. The countryside for miles around is flat and almost featureless. Irrigation and modern farming have transformed the formerly barren plains but in 1848 the countryside was virtually a desert from which the peasants scratched a miserable subsistence.

The inhabitants of Aliwal and the surrounding villages were mostly Sikhs, one of the many sects of Hinduism. Sikhism is essentially a warrior religion and those who embrace it are forbidden to cut their hair, worship idols or smoke tobacco. They were savagely persecuted by the Moghul emperors and learnt to fight hard for their faith. The Sikhs make excellent soldiers and are extremely hard workers. Unfortunately they also possess an inordinate love of intrigue which has led them on more than one occasion to split up into mutually warring clans. Such was their situation at the beginning of the nineteenth century until they were united by a man of genius, Maharajah Runjeet Singh.

He died in 1839 and the unity he had imposed on the Sikhs immediately began to crumble. There was a series of palace revolutions, the army got out of hand, and the Punjab disintegrated into anarchy. These events were watched uneasily by the British who were anxious to avoid a war with the Sikhs. The disasters of the war in Afghanistan were still fresh in people's minds and the East India Company was essentially more concerned with profits than with war. But as the years passed, and the mutinous Sikh regiments in Lahore began clamouring to cross the Sutlej and liberate their co-religionists, the British began to increase the strength of their frontier garrisons. Soon they had concentrated nearly 30,000 men to deal with the expected Sikh invasion, which began on 11 December, 1845 when the main Sikh army crossed the Sutlej.

The 16th Lancers were at Meerut when the orders came for them to march to join General Gough's force on the Sutlej. By Christmas Day, 1845, they had reached Ambala, but it was not until New Year's Day, 1846, that they finally joined up with the main body. By then two fierce battles had been fought, at Mudki and Ferozeshah, and although the British had remained masters of the field their casualties had been heavy. 'On our arrival at the camp ground the stench was horrible,' wrote a sergeant of the 16th. 'A great many were buried within a few yards of our tents. As soon as we had pitched our camp we walked out on the field of battle to view the place and for miles around we could see the dead lying in all directions. At Ferozeshah, about three miles from our tents, the dead were lying in heaps.'

The British and Sikh armies had withdrawn a few miles from each other to lick their wounds. Gough was waiting anxiously for the arrival of his siege artillery which was being dragged slowly by teams of elephants and bullocks all the 200 miles from Delhi. The news that a Sikh force had crossed the Sutlej well to the east worried Gough since this force might intercept the artillery train. A brigade of infantry was sent to drive the Sikhs back across the Sutlej. It was commanded by Major-General Sir Harry Smith.

The brigade marched out for Ludhiana on 17 January, 1846, but shortly thereafter Gough received news that the Sikh strength had increased. Accordingly he dispatched the 16th Lancers to reinforce Smith. This meant forced marches if they were to catch up and it was unseasonably hot. There were no roads and there was little water. The horses' hooves threw up a thick cloud of dust which choked the riders and by the time the cavalry joined up with Smith on 20 January both men and horses were already exhausted. Far behind them toiled the baggage train. It was the custom for regiments to take with them on campaign all the regimental plate, while each officer had a train of ponies and camels to carry his personal belongings. With the baggage went also the sick and wounded. They were carried in doolies – a kind of covered palanquin offering some protection from the sun. This long unwieldy collection straggled along across country, miles behind the main body, guarded against attack only by the few troops which could be spared for the purpose.

Smith allowed the 16th only two hours rest and they were then ordered to lead the advance towards Ludhiana. They were brigaded with two native cavalry regiments under Brigadier-General Charles Cureton, a 16th Lancer himself. He had joined in 1819 after a remarkable career which had begun in 1806 in the militia. In his youth Cureton had run heavily into debt. Soon after joining the army he disappeared. His clothes were found wrapped in a bundle on the beach and it was assumed that he had drowned himself. It later transpired that he had enlisted in the 14th Light Dragoons under the name of Roberts. He fought throughout the Peninsular War in the ranks, and in 1814 was given a commission for gallantry in the field. He then reverted to his own name and five years later joined the 16th Lancers. He never commanded the Regiment but he achieved distinction as a cavalry commander in India and was eventually killed at Ramnagar in 1848.

The advance began soon after midnight on 21 January. As the sun rose higher the day grew hotter. The horses plodded through the sand, led for much of the way by their riders. Any man who fell behind was immediately killed and plundered by the Sikh horsemen hovering round the flanks of the march. The troopers of the 16th mounted many an exhausted infantryman in their saddles or allowed them to hang on by the stirrup leathers. By nightfall the exhausted column arrived in Ludhiana, more or less intact, but the baggage train was less fortunate. It had been under attack throughout the day, many of the sick and wounded had been butchered in their litters, and the greater part of the baggage had been looted. The 16th lost all their regimental plate* and the unfortunate subaltern in charge of the baggage party was placed under arrest by Cureton when he

---

* Many years later a silver-gilt cup was found in a pawnshop in York engraved with the crest of the 16th Lancers. It has always been supposed this cup formed part of the looted plate, but how it found its way to York must remain a mystery.

heard the news. He was to be court-martialled as soon as the opportunity offered but it never did. He charged at Aliwal where he was killed.

The Sikhs withdrew to the Sutlej fords in the vicinity of the village of Aliwal. Early in the morning of 28 January, 1846, the 16th Lancers came up with the enemy. Corporal Cowtan was one of the leading scouts and wrote:

'We came in sight of them about 6 am and formed into line. At this moment the view of the two armies was beautiful indeed – a fine, open, grassy plain, and the enemy in line out of their entrenchments ready to commence; the river in their rear, and in the distance the snowy range of the Himalayas with the sun rising over their tops.'

The Sikh line stretched along a ridge, about a mile long, connecting the villages of Aliwal and Bundri. About 40,000 Sikhs manned the entrenchments dug along this ridge and supporting them were 37 pieces of artillery. On the flanks were the Sikh horse. Smith and his staff took up their position on another ridge, about two miles from the Sikh positions, and watched while the infantry deployed onto the plain to their front. A faint cloud of dust hung over them as they advanced, colours flying in the morning breeze and drummers beating the step. Smith's force totalled about 10,000, of which the 16th Lancers, and the 31st, 50th and 53rd Foot were British. The balance was made up of Company's regiments, two of them being Gurkha battalions. The 16th led the advance, and close behind them were the 31st Foot, 'emulous for the front' as Fortescue wrote later, and charged by Smith with the task of taking by the bayonet the Sikh strong-point of Aliwal. They discharged their mission faithfully but at heavy cost.

The loss of Aliwal jeopardized the rest of the Sikh position since the escape route to the Sutlej fords could now be cut. The Sikh cavalry was thrown into the battle and ordered to recapture the village. A mass of horsemen emerged from behind the ridge at Bundri. Smith, recognizing the danger, sent a galloper to order a squadron of the 16th Lancers and another of the 3rd Bengal Light Cavalry to charge and break up the enemy cavalry, but the 3rd hesitated to obey the order. Captain Bere's squadron set off without them. The light Sikh horses could not stand up to the heavy British chargers, while the British lances out-reached the curved swords of the Sikhs who soon took to their heels. The Sikh cavalry, many thousands strong, had been scattered by barely 100 lancers.

It was a different story with the Sikh artillery and infantry. The guns were exceptionally well served, and the Sikh infantry also fought well. They adopted a triangle formation, and if one side of the triangle was pierced, the other two sides faced inwards and fired indiscriminately at friend and foe. An enemy soldier who fell wounded in this melée had no hope of quarter. He was cut to pieces immediately. As Bere's squadron came galloping back from their successful encounter with the Sikh cavalry they found their way barred by one of these Sikh formations, which greeted them with a volley. Bere did not hesitate but at once charged the Sikhs. He survived with a nasty wound in the face.

According to Corporal Cowtan, who charged with his squadron, 'Sergeant Brown was riding next to me and cleaving everyone down with his sword when his horse was shot under him, and before he reached the ground he received no less than a dozen sabre cuts which, of course, killed him. The killed and wounded in my squadron alone was 42, and after the first charge self-preservation was the grand thing, and the love of life made us look sharp, and their great numbers required all our vigilance. Our lances seemed to paralyse them altogether, and you may be sure we did not give them time to recover themselves.'

The Sikh artillery continued to pound the British and under cover of this fire enemy infantry, supported by a battery of guns, deployed from the ridge as if to attack the right wing of Smith's advancing columns. Two squadrons of the 16th were covering that flank and were ordered to charge and capture the guns. Major Rowland Smyth, who was commanding the regiment that day, at once ordered his trumpeter to sound the 'Trot'. Smyth was a fine swordsman, excellent horseman, and something of a character. He had served a term of imprisonment for killing a man in a duel.* He eventually retired as a

---

* Mr. O'Grady, a Dublin civilian, was accused by Smyth of striking him with a whip while Smyth was driving past O'Grady in a cabriolet.

general, so imprisonment did not affect his military prospects.

When the 16th were within 800 yards of the enemy guns, Smyth turned in his saddle and shouted above the noise: 'Now, 16th, I am going to give the word to charge. Three cheers for the Queen!' The trumpet sounded, the lances came down to the charge, and the whole line surged forward. They rode straight at the guns, the gunners working them until the last possible moment, and then throwing themselves to the ground to avoid the lance points. Just beyond, partially obscured by the dust and smoke, a dense mass of infantry presented the charging squadrons with a wall of swords, shields and bayonets. There was only one thing to do and Smyth did it. He put his horse at the line of Sikhs as if they were a fence out hunting, cleared the first line, galloped through the centre of the triangle of men, collecting a bayonet wound as he went, and jumped out again on the far side. After him came his men.

Sergeant Gould rode in that charge and described it afterwards: 'Down we swept on the guns. Very soon they were in our possession. A more exciting job followed. We had to charge a square of infantry. At them we went, the bullets flying round like a hailstorm. Right in front of us was a big sergeant, Harry Newsome. He was mounted on a grey charger, and with a shout of "Hullo boys, here goes for death or a commission," forced his horse right over the front rank of kneeling men, bristling with bayonets. As Newsome dashed forward he leant over and grasped one of the enemy's standards, but fell from his horse pierced by nineteen bayonet wounds.

'Into the gap made by Newsome we dashed, but they made fearful havoc among us. When we got out the other side of the square our troop had lost both lieutenants, the cornet, troop sergeant-major, and two sergeants. Some of the men shouted, "Bill, you've got command, they're all down." Back we went through the disorganized square, the Sikhs peppering us in all directions. One of the men had both his arms frightfully slashed by a Sikh, who was down under his

---

O'Grady denied any such intention but Smyth pulled him off his horse and horse-whipped him then and there. Whereupon O'Grady called out Smyth and they met next morning in a duel in Phoenix Park. O'Grady was killed.

horse's feet and who made an upward cut at him. We retired to our own line. As we passed the general, he shouted, "Well done, 16th. You have covered yourselves with glory." Then, noticing that no officers were with C Troop, Sir H. Smith enquired, "Where are your officers?" "All down," I replied. "Then", said the general, "go and join the left wing under Captain Bere".'

The 16th paid a heavy price in dead and wounded men and horses. In Smyth's case a bayonet had entered his body just below the waist and had then broken off in the wound. Faint with pain and loss of blood, Smyth nevertheless led the return charge through the broken Sikh ranks, and only after this allowed himself to be taken to the rear. He was in agony, the bayonet having pushed part of his tunic and sword belt into his stomach, but he refused to permit his wound to be dressed until the rest of his wounded had received attention. Six weeks later he was back in the saddle and lived to a ripe old age.

The charge was the culminating point in the battle. The Sikhs began to fall back, the village of Bundri was taken by a bayonet charge, and the British guns were brought forward onto the ridge where formerly the Sikh artillery had been situated. A confused mass of Sikhs fell back towards the river bank, hemmed in by Cureton's cavalry on the flanks, and blasted by Smith's guns at point-blank range. Only a bruised and bleeding rabble reached the far bank, leaving behind all their guns and 3,000 dead. The Sikhs had fought well. Waterloo veterans said they had fought as bravely as the French. The British losses were relatively light but the 16th were not so fortunate. Two officers and 56 N.C.O.s and soldiers were killed; six officers and 77 N.C.O.s and men were wounded, of whom 30 died later. 77 horses were killed, 35 were wounded, and 73 were reported missing. The Regiment had mustered about 500 men and horses at the outset of the battle. By the time the 'Cease Fire' sounded it had lost more than a third of its strength. Little wonder that Captain Pearson, who led a squadron, wrote home to say, 'the 16th Lancers has suffered most severely, much more so than in any battle in the Peninsula or at Waterloo.'

Aliwal was only one of several bloody battles fought during the

First and Second Sikh Wars but it made Sir Harry Smith's reputation as a general. The Duke of Wellington paid him a heartwarming tribute in the House of Lords while Sir Robert Peel moved the motion in his honour in the House of Commons. Public-houses were named after the 'Hero of Aliwal' and the charge of the 16th Lancers captured the public imagination in the same fashion as the charge of the Light Brigade at Balaclava was to do some years later. Today Aliwal has been forgotten except by the Regiment which celebrates the battle every 28 January, but a tall obelisk, slowly crumbling in the Punjab air, still marks the spot where Sir Harry Smith stood with his staff to watch the progress of the battle; and outside the Regimental Quarterguard, where the gong hangs suspended from a triangle of lances, the red and white lance pennons are still carefully crimped. This commemorates the fact that after the last charge had been made at Aliwal the lance pennons were so encrusted with blood that they appeared to be starched and crimped. The tradition has been kept up for 126 years.

Chapter 7
"Goughy"

ON 9 January, 1858, a General Order cancelled the sixty-year-old disbandment of the 5th Royal Irish Regiment of Dragoons. Queen Victoria commanded that the regiment should be restored to its place among the cavalry regiments of the line, and the regiment was formed at Newbridge in Ireland and equipped as lancers. Various reasons are given for the re-raising of the regiment, such as the Queen's recognition of the gallantry of the Irish regiments during the Crimean War, or her pleasure at the loyal welcome she received when visiting Dublin. Whatever the reason, the badge and motto, the Irish harp and crown, and the motto *Quis Separabit* of the old 5th Dragoons was conferred on the regiment. In 1861 its title was changed to the 5th (Royal Irish) Lancers and in 1863 the regiment was sent to India. They remained there until 1874 and on their return home gave the first display of tent-pegging ever seen in Britain. They provided a detachment for the Gordon Relief expedition in 1884, and the following year two squadrons for the Suakin expedition. The battle honour *Suakin 1885* was subsequently given to the regiment.

In 1889 the 16th Lancers were in Aldershot. They were joined by a nineteen-year-old subaltern, Hubert de la Poer Gough, from an Anglo-Irish family. His father and uncle had both won the V.C. in the Indian Mutiny, and his brother also won the V.C. – a remarkable family record. Gough was always known as 'Goughy' in the 16th. In his memoirs* he described life in a cavalry regiment 80 years ago.

'The 16th Lancers were an extravagant regiment, but not more so than many other cavalry regiments. Life was certainly gay . . . We

---

* *Soldiering On* by General Sir Hubert Gough (Arthur Barker), 1954.

had all sorts of dress for different occasions – full dress scarlet tunics, gold lines and girdle; stable jackets, which were short jackets gold-laced and hooked up to the throat; braided blue frock-coats, braided patrol jackets, and, of course, mess dress . . . The N.C.O.s and men were equally particular about their dress . . . As we were scarlet lancers, when our N.C.O.s and men walked out they wore a scarlet monkey-jacket, hooked up to the top of the throat, and a stiff blue collar . . . Their overalls were strapped under the instep, of dark blue, and with two yellow stripes down the sides, and a pair of small well-burnished steel spurs in the heel of their Wellington boots . . . We all wore, when in undress, a small circular forage cap with no peak . . . and we carried a small cane.

'We were all very proud of the regiment and on very good terms with each other . . . There was no squadron organization when I joined, all administration was by troops, of which there were eight in each cavalry regiment. We were expected to know our drill perfectly and to be letter perfect in all words of command . . . practically all our mounted work consisted of drill movement . . . It was strange indeed that the British cavalry – and indeed all arms of the British Army, who had learnt so much under Moore and Wellington in the Peninsular War – had neglected and forgotten all they then knew and practised. After Waterloo the British Army went back to the days of Frederick the Great . . .'

It required the Boer War to jolt the British Army out of its complacency. A succession of Colonial Wars against enemies who lacked modern weapons had made war seem easy. The Boers were soon to demonstrate its difficulties. Gough was a young officer who studied his profession. There was another young 16th Lancer who held the same views about soldiering but he moved on just before Gough joined.

He was William Robertson, son of a village postman, who enlisted into the regiment in 1877. A man of strong and determined character, he rose rapidly through the ranks to troop sergeant-major. He never lost an opportunity to improve his education and military knowledge, and he was fortunate in obtaining the support of several of his officers. In 1888 he was commissioned into the 3rd Dragoon Guards, although he was entirely lacking in private means, and by 1910 he was a Major-General and Commandant of the Staff College. In 1915 he became Chief of the Imperial General Staff and he ended his career as a Field-Marshal and a Baronet. He died in 1933 and must remain an inspiration to every young man joining the 16th Lancers. In an age when privilege counted, when some kind of private means seemed essential for every cavalry officer, this remarkable self-educated man rose through every rank in the army to reach the highest peak. His son, Lord Robertson of Oakridge, followed him into the army and ended his military career as Adjutant-General.

In 1897 the 5th Lancers were in India. Their commanding officer was Lieutenant-Colonel 'Jabber' Scott Chisolme, an outstanding officer who was killed in South Africa. At the conclusion of the training season Chisolme wheeled the regiment into line and sounded the 'Officers' call. After the officers had galloped out to him, he said, 'Gentlemen, I have called you out to look at such a regiment of cavalry as you are unlikely ever to see again. Turn about and look at the Regiment;' and as the junior subaltern present was later to record, 'I knew the Colonel was right as with pride we gazed on that long line of five strong squadrons standing motionless under the Indian sun, not a horse out of place . . .' Indeed, the 5th Lancers left India in 1898 for Natal with a reputation second to none.

They were stationed at Maritzburg when war with the Boers broke out in October, 1899. On 21 October they charged at Elandslaagte and scattered the Boers, but lost their colonel, killed while binding up a wounded trooper. Two weeks later they were besieged in Ladysmith with the rest of General White's Natal force. The siege lasted four months and rations were short. 'For dinner today,' wrote a diarist on 2 February, 'we had chevril (a horse bovril) and the haunch of a mule. No doubt one could manage it if the meat and soup were good instead of being tainted by the hot weather and the plague of flies.'

Ladysmith was relieved by a mounted column containing a squadron of the Imperial Light Horse and the Natal Carabineers, volunteers all. Hubert Gough was serving with this force and was the first officer into Ladysmith. There he was greeted by Sir George White, pale and drawn from strain but otherwise in full control of his emotions. 'Hullo, Hubert, how are you?' was all he said. Later Gough met his brother John, who had served throughout the siege with the Rifle Brigade. He too greeted Hubert in matter-of-fact fashion – 'How fat you've got!' Gough himself was very moved by the occasion but extremely irritated by a young officer, Winston Churchill, who had attached himself to Gough's column. 'He was fussy!' he complained.

The 16th Lancers did not arrive in South Africa until early in 1900 – also from India. They took part in the Relief of Kimberley, the battles of Paardeberg and Diamond Hill, and did not return to England until November, 1904. The 5th Lancers returned from South Africa in October, 1902. The South African War was one of those traumatic experiences which occur every so often in the history of the British Army. At the outset the army's tactics were out-of-date, its generals too old and set in their ways, and the enemy was underrated. The result was a succession of disasters. Later, learning from experience, tactics improved, new ideas were introduced, and a much fitter and more battle-worthy army was the result. Both the 16th and the 5th Lancers benefitted from all this and returned from South Africa in better shape to compete with modern war.

King Alfonso XIII of Spain was appointed Colonel-in-Chief of the 16th Lancers in 1905 and took a keen interest in the Regiment for the rest of his life. He was married in 1906 to Princess Victoria of Battenberg in Madrid and four officers of the Regiment attended the ceremony. A bomb was thrown at the King and his bride as they drove ceremonially through the streets. Fortunately neither was injured, but the 16th Lancers contingent hastened to the King's side and escorted the Royal Carriage back to the Palace.

The Regiment's close connection with King Alfonso has resulted in two interesting traditions, one certainly founded on fact, while the other may be apocryphal. The first tradition is that the Spanish Royal Anthem is played in mess on Dinner Nights immediately before the National Anthem.* The second tradition relates to the Regiment's practice of wearing the cross strap of their Sam Browne

---

* Although the National Anthem is played in Mess on Dinner Nights, the officers do not drink the Loyal Toast nor stand for the Anthem. The origin of this custom is shrouded in mystery!

belts the wrong way round, i.e., fastening at the back rather than in the front. Tradition has it that King Alfonso was inspecting the Regiment on Salisbury Plain in the early 1920's. The Commanding Officer was Lieutenant-Colonel (later Major-General) Geoffrey Brooke, an Anglo-Irishman in the Gough tradition, a brilliant horseman, an excellent writer, and the complete *beau sabreur*. When the King, wearing service dress uniform, appeared in the mess prior to the parade, Brooke noticed that the King was wearing his cross strap the wrong way round. He could hardly tell the King that he was improperly dressed! Brooke therefore told all the officers to fasten their cross belts in the same manner as the King's, and they have been worn in this fashion ever since.

The 5th Lancers, on their return from South Africa in 1902, received a new commanding officer. He was Lieutenant-Colonel E. H. H. Allenby of the 6th Iniskilling Dragoons, who was to achieve great distinction during the Great War and after. Allenby was a big man in every sense of the word but he had an unfortunate tendency to bully people which earned him the nickname of 'The Bull.' Gough served Allenby later as his principal staff officer, and although he liked him, compared him unfavourably with Haig. He paid too much attention to detail and was a slave to regulations; he could be mentally lazy, and never seemed sure of exactly what he wanted. He certainly did not display these qualities, or lack of qualities, twelve years later in Palestine where he showed himself to be a master of the art of mobile warfare.

Hubert Gough became commanding officer of the 16th Lancers in 1907 at the age of 37. Not only had he attended the Staff College, which was not particularly encouraged in the 16th Lancers in those days, but he had gone back there as an instructor and was clearly an officer with a future. According to Brigadier-General Beddington, another brilliant officer who served under him, Gough was an inspiring commanding officer. His interests lay in training, not in the parade ground, and he was a first-class tactician. 'By the time Goughy's command was over,' wrote Beddington, 'we were just about the best trained cavalry regiment in the Army.' Gough was equally keen on sport and insisted that his officers kept fit. He was a good disciplinarian but he had a mind of his own and was not afraid of speaking out when he disagreed with his superior officers. He was brave both physically and morally, the latter being the rarer quality among officers with ambitions for the future.

It was this moral courage which got Gough into trouble in March, 1914. At the time he was commanding the 3rd Cavalry Brigade, consisting of the 16th Lancers, 4th Hussars, and 5th Lancers at the Curragh in Ireland. Ireland was in the throes of a crisis, caused on this occasion by the refusal of the Protestants in Ulster to accept Home Rule. Led by Sir Edward Carson and James Craig, they had said they would fight rather than accept, but Gough said that the British soldiers serving in Ireland were not much interested in what they conceived to be a political quarrel. There have been many versions of what has since been called the 'Curragh Mutiny' (which it certainly was not), but Beddington's is as good as any. He had been lecturing to the Staffordshire Yeomanry and was returning to the 16th Lancers when he was cheered by the porters on Crewe railway station. He enquired the reason and was told: 'Your regiment refused to go and fight Ulster; here's luck to you all.' Beddington goes on to say:

'I called at Barracks on my way home and found that on the Friday (20 March), General Paget, G.O.C. Irish Command, had sent for Goughy and others and told them that they and their officers had the choice of resigning their Commissions (in fact being dismissed) or of most probably engaging in active operations against Ulster. Any officer with an Ulster domicile would be allowed to disappear. Paget required an answer that evening. Goughy saw the 5th Lancers in Dublin that morning, and the other officers of the Brigade in our mess that afternoon, and the vast majority (all in the 16th Lancers) decided for resignation or dismissal, and reported to Dublin that evening that of the officers on duty 59 out of 71 preferred dismissal. The next day General Paget went to The Curragh to see the officers of the Brigade stationed there, and endeavoured to get them to reconsider their decision; there was some weakening but very little. Gough, and the colonels of 5th and 16th Lancers were ordered to report to the War Office the next day, Sunday . . .'

Seldom in the long and sorry history of England's relationship with Ireland has anything been worse handled. Gough and his officers acted throughout with calmness and perfect discipline, but the business was grossly mishandled at the highest level. It became a political sensation, leading to the resignations of the Secretary of State for War (Seely) and the C.I.G.S. (Sir John French). General Paget, who was responsible for the nonsense in the first place, was retired shortly afterwards. It could have damaged the careers of many other officers, including Gough's, but the Great War broke out on 4 August, 1914, and men's minds were focussed on sterner issues than the apparently insoluble problems of Ireland.

Gough took his cavalry brigade to France on 15 August. One week later one of his batteries – E Battery R.H.A. – fired the first British shell of the war at the Germans as they advanced at Mons. Millions of shells were to be fired before the war was over. The 16th Lancers suffered their first casualties on 22 August when Lieutenant Tempest-Hicks' Troop came upon German infantry when reconnoitring at Peronnes, near Mons. They were concealed among the recently cut wheat stooks and opened heavy fire on the 16th Lancers patrol which immediately charged them. Tempest-Hicks' horse was shot under him, and two other horses were killed and a soldier was wounded, but covering fire from E Battery enabled the patrol to withdraw safely. Gough wrote of Tempest-Hicks that he was 'one of our most gallant young officers.' He fought throughout the rest of the war, mostly in the trenches, but was unlucky to be killed shortly before the Armistice by a chance shell some distance behind the lines.

The Great War provided little opportunity for the cavalry to distinguish themselves in 'shock action'. It was essentially an infantryman's and artilleryman's war – an affair of mud and blood, trenches and barbed wire, raid and counter-raid, poison gas, massed attacks, bombardment and counter-bombardment. Although the cavalry kept their horses, they had little opportunity to use them. The 16th and 5th Lancers covered the retreat of the B.E.F. from Mons, a period which was as confusing as it was exhausting, and on one occasion during the retreat two Troops of the 5th Lancers were cut off and surrounded. They put up such a stout resistance before surrendering to the enemy that almost every man was wounded. One trooper, Kay, became separated from the rest during the retreat. He was called upon to surrender whereupon he took up a position in a carriage where he defended himself until killed by a volley. The local villagers declared he killed six German officers before being killed himself and his conduct impressed them so much that they buried him where he fell and erected a cross above the grave.

By the middle of October, 1914, the German advance on Paris had been halted and thrown back. The front had begun to stabilize and trench warfare began from the Channel coast right across France to Switzerland. The 16th Lancers went into the trenches as infantry on 20 October near Armentières, and the 5th Lancers soon followed them. The latter, with an Indian regiment in support, were almost decimated at Hollebecke by artillery bombardment. All the British officers in the Indian regiment were either killed or wounded and the Indians began to give ground. Trooper Colgreave, who spoke Hindustani, rallied a number of them and led them back to the trenches where they fought bravely. Colgreave was awarded the D.C.M.

During the winter months of 1914-15 the regular regiments were bled white defending the notorious Ypres salient. Both the 16th and 5th Lancers played their part in holding the line in some of the most bitter fighting of the war. The trenches were waterlogged and for much of the time men fought up to their knees in water. Even the rest areas were within reach of the enemy guns but nevertheless there were football matches and concerts. Major Barry of the 5th Lancers returned from leave bringing with him several couple of fox-hounds. The French refused at first to release them, arguing that no sport was allowed during wartime, but by fair means or foul they were got to the regiment near Ypres. For several weeks they provided good sport, until the French came to hear about them and insisted that hunting had to stop.

Many writers give the impression that for much of the war in France and Flanders the cavalry hung about in the back areas waiting for the chance to break through – a chance that never came until the very end. This is not a true picture. For much of the time both the

16th and 5th Lancers took their turn in the trenches as infantry. The 5th Lancers particularly distinguished themselves in the defence of Guillemont Farm in June 1917. Although the trenches were deep and well-sited, there was only one dug-out in which to take refuge when the Germans opened their bombardment. At one time it seemed certain that the position would have to be abandoned but the 5th Lancers hung on. They lost 16 men killed and 24 wounded in this action, and were awarded for their gallantry two M.C.s, one D.C.M., and four M.M.s. In December of the same year a squadron of the 5th Lancers again distinguished itself in the defence of Bourlon Wood. During this action Private Clare won a posthumous V.C. He was a stretcher bearer who dressed wounds and helped evacuate wounded men under intense machine-gun and artillery fire. 'At one stage, when all the garrison of a detached post, which was lying out in the open . . . had become casualties, he crossed the intervening space, which was continuously swept by heavy rifle and machine gun fire, and, having dressed all the cases, manned the post single-handed until a relief could be sent.' He then carried a badly wounded man through intense fire to the dressing station, where he learnt that the enemy was using gas shells in the vicinity and the prevailing wind might blow the gas over the 5th Lancers position. On his own initiative he personally warned every company post of the danger, all the time under heavy fire. This very brave man was eventually killed and the V.C. awarded to him was the first to be given to a member of the 2nd Cavalry Division.

Although the cavalry's role was restricted, there were occasions when the horse came into its own. One such occasion was the charge of Lord Strathcona's Horse at Moreuil Wood on 30 March 1918 when the Canadian Cavalry Brigade counter-attacked the German advance and prevented them from breaking through. The Canadians suffered heavy casualties, as might be expected from horsed soldiers attacking infantry, but at a critical moment they were supported by the 3rd Cavalry Brigade. In the lead were the 16th Lancers, commanded by Lieutenant-Colonel Geoffrey Brooke. He had been Brigade Major of the Canadian Cavalry Brigade before his appointment to command the 16th Lancers and the Canadians regarded him as one of themselves. An Anglo-Irishman, and quite imperturbable under fire, Brooke could see that German reinforcements were infiltrating into Moreuil Wood and extending their right flank outside the wood. He led the 16th Lancers in a dismounted attack against the southern corner of the wood, while the 4th Hussars operated mounted on his flank. The attack was successful but there was bitter fighting inside the wood.

Brooke had the reputation of being completely unmoved by enemy fire but not every soldier could hope to be quite so stout-hearted. In his account of the attack, he wrote:

'I had just passed the word down the line to advance when a soldier, who had temporarily lost his nerve, started to run back. I had a large pair of wirecutters which I hurled at him and hit him on the knee. This may have restored his equanimity as he then carried on – or it more likely may have been due to the remark of an old soldier seeing the German machine guns ripping up the grass in front of us, "God," he said, "it reminds me of old Nobby cutting up the billiard table."'

Just before Moreuil Wood a composite cavalry regiment was formed from the 3rd Cavalry Brigade and placed under the command of Lieutenant-Colonel Geoffrey Brooke. Its task was to hold the important Porquericourt Ridge which the Germans were threatening to occupy. Brooke led his regiment at full gallop to reach the ridge before the advancing Germans but arrived only to find that he had been beaten to it. All but the highest part of the ridge had been seized by the enemy. The highest point was 'No Man's Land' and the British were the first to get there, and having done so, they held it against all the German attempts to dislodge them. In Harvey's *History of the 5th Lancers*, he writes: 'I am told that there has been no more splendid sight in this war than this wild rush of cavalry across fields and through villages to gain the coveted ground.'

Such opportunities were, however, rare. The cavalry may have kept their horses but in battle they were more likely to be called upon to fight on foot. They may not have suffered as severely as the infantry, but the 16th Lancers lost 22 officers and 157 N.C.O.s and Troopers killed or died, and the 5th Lancers lost eight officers and 98 N.C.O.s and Troopers. Both regiments fought throughout the war in France and Flanders, for most of the time in the 2nd Cavalry Division. They were at Ypres where the old Regular Army went down fighting. They were at St Julien and Bellewaarde in 1915, in reserve for the Somme in 1916, at Arras, the Scarpe and Cambrai in 1917, and at Amiens in August, 1918, when the Germans first began to crack. Both the 16th and the 5th took part in the Pursuit to Mons, retracing the steps they had taken four years previously, and in November, 1918, the 5th Lancers, attached to the Canadian Corps, were the first British troops to re-enter Mons. Few who had ridden out of Mons were with them when they rode back.

The 'Scarlet Lancer' to reach the greatest heights during the Great War was Goughy. And when his luck ran out – and no General can hope to succeed without luck – he fell the farthest. He was a Brigadier-General in 1914 at the age of 44; he was a Major-General by the following April as G.O.C. 7th Infantry Division; by July he was a Lieutenant-General as G.O.C. 1st Corps. In May 1916, while not yet 46, he was selected by Field-Marshal Sir Douglas Haig to command the Reserve Army, which subsequently became Fifth Army. It was remarkably quick advancement when one considers that Montgomery was 55 when he took command of Eighth Army, and Slim 53 when he took over XIV Army. It was Gough's Fifth Army, undermanned and overstretched, which bore the brunt of the great German offensive of March, 1918. Gough fought an able delaying action which in the end stopped the Germans, but at a cost. His rapid rise had inevitably made him enemies; he was thought to be inconsiderate of men's lives, and insufficiently ruthless with his staff. A scapegoat was demanded by the Prime Minister (Lloyd George), and people conveniently forgot that Gough had never ceased to warn that he had insufficient troops to hold his sector securely. He was sacked without hope of redress on 28 March, 1918, and being the man he was, he forbore to add to the terrible responsibilities of his Commander-in-Chief, Sir Douglas Haig. He returned home to a succession of interesting, but minor, appointments and retired from the army in 1920. Throughout the rest of his long life (he died in 1963, aged 93) he never ceased to campaign for the removal of the slur on the record of Fifth Army but he never said a word on his own behalf. He was fully exonerated when the Official History was published in 1936 and in 1937 he was made G.C.B. He was Colonel of the Regiment from 1936–43 and is certainly one of the most outstanding and attractive officers ever to have served with the 'Scarlet Lancers'.

Chapter 8
16th/5th Lancers

THE Great War of 1914–18 saw the end of the old order of things. Nothing was quite the same again. Of the thirty-seven officers on the regimental list in 1914, only fifteen were still serving with the 16th Lancers in 1919. It was much the same with the 5th Lancers when they embarked for India at the end of that year. The minds of some cavalry officers may have remained set in the pre-war mould but the whole nature of war had changed. As usual, after all Britain's wars, the first cry was to reduce the size of the army, and thereafter to spend as little money on it as could be contrived. This policy directly affected the 5th Lancers who were stationed in Peshawar in March, 1921. Their commanding officer (Lieutenant-Colonel Cape) was attending a race-meeting when an acquaintance enquired whether Cape had seen the latest Reuter telegrams, adding, 'They're going to disband your regiment!' The astonished colonel hastened to obtain a copy and by this means, an ordinary press telegram, discovered that the 5th Lancers were once more to be struck out of the *Army List*. It was to be two months before official notification arrived from the War Office, offering the 5th Lancers the choice between conversion to a battalion of the Royal Tank Corps or disbandment. The officers unanimously chose disbandment.

There was a hard fight in Parliament to get the decision rescinded but the Army Council stood firm. Officers and soldiers left the service or were posted to other regiments. By the end of 1921 the 5th Lancers had ceased to exist. Then in April, 1922, there was a change of heart. Army Order 133 announced the amalgamation of certain cavalry regiments, and among them the 16th Lancers with the 5th. The combined regiment was to be titled the 16th/5th Lancers because the 5th Lancers, when they were reformed in 1858, were placed junior to the 16th in the *Army List*. This in itself did not make for a happy union but even worse was the fact that most of the original 5th Lancers were unwilling to leave their new regiments and join the amalgamated one. It proved necessary in the end to draft 99 16th Lancers into D Squadron (the 5th Lancers squadron) to bring it up to strength. The two regiments had served together for many years but were different in character and it was hardly a happy marriage. The cap and collar badges of the 16th Lancers were retained, as also various other idiosyncracies of dress. 5th Lancer buttons replaced 16th Lancer ones and all full rank N.C.O.s wore the Irish Harp on their chevrons. The two regimental marches were blended into one. 5th Lancer officers continued to wear their own badges and uniform for some years, including having their field boots made by a different bootmaker than the 16th, but these differences gradually died away. Inevitably, however, the 16th Lancers became the dominant partner and the Old Comrades' Associations of the two former regiments remained separate.

As happens sometimes in ordinary human relationships an unhappy union was transformed by an event outside the control of either party. The Second World War fused the two regiments into one, although by 1939 few of those serving in the 16th/5th Lancers could be bothered with the quarrels and arguments of 1922. However it did result in the amalgamation of the Old Comrades' Associations and this put the final seal on the union.

The Regiment spent many of the inter-war years in Egypt and India. In 1924 they escorted Lord Allenby,* High Commissioner in Cairo, when he drove through the streets to deliver the British ultimatum to the Egyptian Prime Minister after the assassination of Sir Lee Stacke, Governor-General of the Sudan. But for the most part the Regiment was employed in routine duties, earning for themselves a high reputation at a time when the Army was being starved of money. The nation as a whole had lost interest in the Army, as had happened before after Waterloo, and the Army in its turn lost touch with the nation. Training was unrealistic, undue time

---

* Field-Marshal the Viscount Allenby was joint Colonel of the Regiment at the time.

was spent on Tattoos, and officers and soldiers alike tended to concentrate on sport. The British had invented the tank, and had in General Fuller and Captain Liddell Hart the foremost exponents of mechanized warfare, but far too much time and energy was wasted in the futile controversy of horse versus tank. It may seem incredible today that there were soldiers who believed that the horse still had a place on the battlefield, but this was the case and many of them wielded great influence. It was an emotional issue and the Treasury sided with emotion because mechanization would be expensive. In 1927 the Army Estimates provided £607,000 for forage and only £72,000 for petrol. In 1936 Parliament was informed of the intention to mechanize eight cavalry regiments. 'It is like asking a great musical performer to throw away his violin and devote himself in future to the gramophone,' apologized the Secretary of State for War.

The 16th/5th Lancers had to wait until 1940 to exchange their horses for tanks. The Regiment was at Risalpur (now Pakistan) when war broke out and the last mounted parade on horses was held there in November. They returned to England at the end of the year and in May 1940 received their first tanks – they were five Great War veterans of dubious mechanical efficiency. The days of *L'Arme Blanche* were finished but it would be some time before every 16th/5th Lancer could reconcile himself to the fact. Some of them never did and took themselves off to employment more congenial than tanks. However, time was to show that the cavalry were able to adapt themselves to their new role, but inevitably there were teething troubles at first. The Regiment is greatly indebted to Lieutenant-Colonel Macintyre, who was the commanding officer during the crucial period of changeover from horses to tanks.

On 14 November, 1942, the Regiment sailed from the Clyde for North Africa as part of 6th Armoured Division. Lieutenant-Colonel Babington was in command.* The first three months were spent between Teboursock and Medjez el Bab – 'Three months of rain and mud, dreary days in tank bivvies by the side of our Valentines

---

* Colonel Babington's father, Lieutenant-General Sir James Babington, commanded the Regiment from 1892–96, and was Colonel of the Regiment from 1909–36.

and Crusaders. Days spent in brewing up our food into all forms of messes and hashes, and in drying out our clothes.' The first tank casualty was a Valentine, brewed-up by an enemy plane through the thickest part of its armour. The news after Christmas that the Regiment was to be issued with brand-new Shermans was enthusiastically received but before the new tanks could be made operational orders came for action. The Regiment therefore fought its first major engagement in obsolescent Crusaders and Valentines.

If the first three months in North Africa had been boring, there was excitement enough to follow. The 16th/5th Lancers were heavily involved in three major battles – Fondouk, Kournine, and the final battle for Tunis. There was also a fierce action at Bordj on 11 April, 1943, when the regiment actually 'charged' the enemy. Bordj was unusual since the Regiment operated as a whole instead of by squadrons. 'A great clang, and the turret was full of flames,' says one account. 'It seemed only a fraction of a second that I was struggling to open the front lid of the turret, but in that time the gunner had slipped up from his seat in front of me, pushed open the lid and clambered out. It was getting unbearably hot. I remember thinking that I must get out quickly before the heat took my strength. With a heave I made it and jumped from the top of the turret to the ground, rolling over and over to put my clothes out.' Although burnt, wounded and badly shocked, he lived to fight another day, but being 'brewed up' by enemy gunfire was always the armoured soldier's nightmare.

It was at Bordj that Lieutenant (later Colonel) Frank Watson was awarded the M.C. for gallantry and leadership when in command of the Reconnaissance Troop. He was later to win the D.S.O. in Italy. He was in the Robertson tradition since he joined as a Trooper and rose through every rank to Colonel – an outstanding record. Fortunately by the time the Regiment was heavily involved in North Africa they had disposed of their Crusader and Valentine tanks and received the American Sherman tank in their place. Their faith in the former had received a nasty jolt at Sbeitla when a 'Tiger' tank had knocked out three Crusaders at a range of almost 2,500 yards – 'It was a nasty shock!' But the Sherman, mechanically reliable with reasonable protection and a good gun, was an altogether different proposition. It served the Regiment well for the rest of the war, and was indeed performing yeoman service in other armies 25 years after the war.

Although the German situation in Tunisia seemed hopeless, they fought fiercely until the very end, and took their toll of the 16th/5th Lancers. The Regiment buried three officers and 61 N.C.O.s and Troopers in North Africa but their sacrifice was not in vain. On 13 May, 1943, Field-Marshal Alexander was able to make his now famous signal to the Prime Minister:

'Sir, it is my duty to report that the Tunisian campaign is over. All enemy resistance has ceased. We are masters of the North African shores.'

So far as the Regiment was concerned the campaign had put the seal on two years of training in Britain, followed by experience of battle as an armoured regiment. They were now seasoned tank soldiers and part of a well-trained team – the 26th Armoured Brigade of the 6th Armoured Division, whose mailed fist insignia they wore on the sleeves of their battle dress blouses. With them in the 26th Armoured Brigade were the 17th/21st Lancers and the 2nd Lothians and Border Horse. Eight months after the occupation of Tunis the 16th/5th Lancers were in action again – this time in Italy where they were to experience some of the hardest campaigning in the long history of the Regiment.

'During the seventeen months in which we were engaged in Italy,' writes the regimental chronicler in *Scarlet and Green*, 'I do not suppose it was for more than five that we were employed in conditions as any armoured regiment enjoys – that is, with room to manoeuvre, in a normal countryside, squadrons supporting squadrons, and together (perhaps with infantry) hustling the enemy with the enormous fire power that we now have in our tanks. Instead, you would find us in vineyards and olive groves, a tank here, and a tank there; a squadron split up on an infantry brigade front, grappling with mud and with mine-fields, with a field of view not very much longer than the length of our guns. That was Minterno. Again you might have found us perched in our tanks on the very heights of the Appennine Mountains looping our way up a zigzag lane hoping to force a way through some suspected defile and to debouch behind the enemy into the Po Valley. At another time we became infantry. Deserting our tanks, again in mountains, we undertook six weeks of vigorous patrolling on our feet' – and in deep snow!

For the majority of the time it was tough, hard slogging from one river line to the next, with rain, mud and snow as much the enemy as the Germans. At Minterno, and the crossings of the River Garigliano which preceded that battle, the task was to help the infantry forward. 'It was nothing like tank country: with close cultivation, high hedges, olive trees and grape vines, you could seldom see for more than a few dozen yards in front. Tanks got bogged in bottomless ploughs and the enemy were sited with anti-tank guns too well placed to allow risks to be taken . . . Sometimes we had one squadron up with two back, and sometimes the other way round. Those that were back trained with the infantry, practised co-operation with them and generally got to know their wants and ways. Then they went up to the line. It was a laborious and tiring business. Each squadron would stay up about ten days and then go back for a breather, a bath at the mobile showers, a trip or two into the city (Naples) to see an E.N.S.A. show and more training.'

The abortive battle for Cassino followed, when the Regiment waited eagerly for the opportunity to break through, an opportunity which never came despite the gallant attempts by the infantry to break the enemy resistance. By May the Regiment had rejoined the 6th Armoured Division, now concentrated in Italy, and the long-awaited final assault on Cassino took place. On the 15 May the 16th/5th Lancers began to move forward, with very limited objectives in view of the close country and the debris of the battle-field, but they were moving in the right direction. It was on that same day that the Regiment lost its popular and gallant commanding officer, Lieutenant-Colonel John Loveday, who was killed while arranging an attack to consolidate the regiment's position. Loveday's loss was deeply felt in the 16th/5th Lancers. He had had a miraculous escape in Tunisia when captured by the Germans. Brought before a firing squad he had fallen sideways simultaneously with fire being opened and survived unhurt, subsequently rejoining his own lines.

Had he lived it is certain that he would have achieved great distinction in the Army.

The advance up the Liri Valley followed the breakthrough from Cassino, and it was during this advance that A Squadron particularly distinguished itself in the capture of Piumarola on 17 May, 1944. The fight for Rome was now in full swing. After crossing the River Melfa the country was more open and the tanks came into their own. 'We went like scalded cats, advancing so swiftly that we all but ran off our maps; in fact, at one place the Intelligence Officer had to stand by the side of the road handing out maps of the country ahead to each tank as it passed.' Rome fell on 4 June but the advance continued. Perugia fell on 20 June. 'There we stayed ten days – ten days to reorganize, rest and do repairs. It was a very pleasant town and we got ourselves some splendid billets – a girls' school. The girls, worse luck, had gone!' Perugia was the first real break since Cassino and it was to be the last for many weeks to come. Florence was to be the next objective and it seemed almost round the corner – three weeks fighting at the most – but it turned out to be nearer eight weeks. There was fighting all the way to Florence, including a hard battle for Arezzo, which the Regiment entered on 10 July. Five days previously it had lost yet another commanding officer, Lieutenant-Colonel Gundry, who had been wounded by a shell fragment. Lieutenant-Colonel Nicholson, himself a 16th/5th Lancer, who was commanding the 2nd Lothians and Border Horse, returned to the Regiment to take command.

All replenishment and tank maintenance had to be done after dark because 'even a flicker of light was liable to attract shell fire.' As Florence grew nearer, the country became more difficult and mountainous. Squadrons became increasingly expert in employing indirect fire to harass the enemy and help the infantry forward. Booby traps and mines took their toll of men and machines on the narrow mountain tracks and progress was slow as the 'Sappers' repaired the frequent 'blows' made by the Germans as they withdrew deeper into the mountains. This was the worst tank country in Italy, and that is saying a great deal. There was a short break from mountaineering just before Christmas, 1944, when the Regiment was suddenly moved to Osimo on the Adriatic coast, the rumour being that it was to embark for Greece. But on Christmas Eve the order was cancelled and by 14th January the 16th/5th Lancers were back in the mountains above Florence, this time to operate as infantry. For nearly six weeks they held part of the line in the area of Casola Valsenio where the main problem was getting supplies and ammunition to the forward positions. Mules took the place of trucks and jeeps and brought back nostalgic memories to those of the 'old and bold' who could remember the days of the horse.

They were relieved in the line and moved to Pesaro on the east coast in March. There they were reissued with tanks and began training for what was to turn out to be the final offensive of the war in Italy. Working in close liaison with the 1st battalion of the 60th Rifles each squadron and company was welded together as a battle group. The 8th Army attack was launched on the evening of 9 April, 1945, but it was not until 19 April that the Regiment and the 60th moved forward. 'For five days, leap-frogging in turn with the other regiments of the Brigade, we buffeted our way through the remaining enemy lines and lashed into his attempts to reorganize himself in front of the Po. We had wonderful support from the 60th, from our guns and from the air, and we were able as a Brigade so to bounce our way over the various river and canal lines that the enemy might otherwise have made good, that he was utterly beaten and disorganized south of the River Po and was forced to abandon practically all his equipment and vehicles.'

It was all over on 2 May, 1945, and when the final German surrender came in Italy the 16th/5th Lancers were 'harboured round some farm buildings a few hundred yards from the Po.' It was a long way from Risalpur on the Punjab plains where they had started from, and the Sherman tank with 76 mm and 105 mm guns had little in common with horse and lance. Lieutenant-Colonel Smyly had led the Regiment to victory and he was later to be Colonel of the Regiment from 1959–69. It was in some way symbolic of the Regiment's success in mastering new techniques that its commanding officer when the war ended happened to be an exceptionally good horseman, as well as a most capable soldier.

The German soldier had fought as well in Italy as he had done in North Africa. From the beginning of the campaign until the end he was quick to seize the fleeting opportunity in attack, and tenacious in defence. If war be a hard school, the German soldier proved himself to be a tough teacher. The 16th/5th Lancers buried 32 of their comrades in Italian soil (four of them officers), and of all the many Battle Honours won in their long history, *Italy 1944–45* was harder earned than most.

Chapter 9
The Uneasy Peace

A LONG war has profound effects on a regiment. It is never the same at the end as it was when the war began. Many have fallen by the wayside, killed, wounded, or gone elsewhere. Equipment and tactics may have changed, new ideas been adopted, and there may even have been sociological change as well. This was certainly the case during and after the Second World War. The British Empire has ceased to exist and our place in the world has altered dramatically. The Army has gradually been reduced in strength until today it totals little more than 150,000. There are few overseas garrisons left, apart from Germany, and military equipment has become increasingly complex. Compare, for example, the Chieftain tank with the Great War veterans issued to the Regiment when it was first mechanized in 1940. The disappearance of the empire has been marked by many small wars, but the only campaign in which the Regiment was involved was in Aden. Yet it has spent most of the past 25 years overseas – in Austria, Germany (four times), Egypt, Cyrenaica, Tripoli, Aden, the Persian Gulf, and Hong Kong. For most of the time it has been equipped with tanks as an armoured regiment, but from 1945–51 it was organized as an armoured car regiment, and in 1971 again reverted to the reconnaissance role. From 1957–59 it served as a Basic Training regiment for the Royal Armoured Corps. Whatever its role, it has endeavoured to make a success of it.

* * * *

On 19 March, 1959, the Regiment paraded at Buckingham Palace to receive its Guidon from Her Majesty The Queen. The Queen had become Colonel-in-Chief of the Regiment on 21 April, 1947, her 21st birthday, when she was The Princess Elizabeth. Although it is usual when the Sovereign ascends the throne for previous appointments to lapse, the Regiment were delighted to learn at the time of the Queen's coronation that Her Majesty was pleased to retain the Colonelcy-in-Chief. She also conferred a new title on the Regiment – 16th/5th *The Queen's Royal Lancers* – which appeared in the *Army List* for the first time in August 1954. In the same year a new collar badge was taken into use. This combined the badges of the 16th and 5th Lancers and is known as *The Queen's Badge*.

The Presentation of the Guidon was an historic event. Guidons are swallow-tailed flags with the Battle Honours embroidered on them and are consecrated before being taken into use. They were abolished for Lancer regiments in 1834 and were not re-introduced until after the Second World War. In the interim the regiment's Battle Honours were embroidered on drum banners which were draped round the mounted kettle-drums. Before the Queen presented the Guidon the 16th Lancers' drum-banners, draped round the 5th Lancers' silver kettle-drums, were ceremonially marched off parade to the tune of 'Auld Lang Syne'. The mounted escort wore the full-dress uniforms of the 16th and 5th Lancers. That same night the officers gave a dinner and ball to mark the occasion, and also to commemorate the fact that 1959 was the bicentenary of the 16th Lancers. The Queen attended, as also did Their Royal Highnesses The Duke and Duchess of Gloucester, the Duke having served with the Regiment in 1931–32 when his own regiment was overseas. 1959 was indeed an *annus mirabilis* for the Regiment, since in April the Football XI also made history by being the first cavalry team ever to win the Army Cup.

* * * *

An unusual feature of the post-war period has been the number of occasions when the regiment has been split up with one or more sabre squadrons serving away from regimental headquarters, a state of affairs more welcome to the squadrons than to the unfortunate commanding officers who have had to maintain control. 1963–65 was probably the most difficult time when R.H.Q. and one squadron was in Aden, one squadron was in the Persian Gulf (half in Bahrein and the other half embarked in an L.C.T. at immediate operational readiness), while the third squadron was in Hong Kong. The unfortunate commanding officer (Lieutenant-Colonel Holland) covered many thousands of miles by air while visiting his far-flung squadrons. The Regiment was also involved in the Radfan operations in the Aden hinterland. These began in January, 1964, and continued throughout the year. The Reconnaissance Troop and two Tank Troops were involved at various times and many members of the Regiment became eligible for the General Service Medal with the *Radfan* and *South Arabia* clasps.*

After a spell in Tidworth the Regiment went out to Germany in 1968 for a highly successful tour, culminating in its being chosen to represent B.A.O.R. in the annual Canada Cup competition. The trophy was presented in 1963 for tank gunnery and is competed for by the member nations of N.A.T.O. In 1970 the other competitors were the German and Canadian armies, the former equipped with the new Leopard tank, while the Regiment had the Chieftain. The competition is designed to test every aspect of tank gunnery and to win it demands a high degree of skill and efficiency. In the event the Regiment won handsomely over the Germans, with the Canadians third. Indeed, it can be said to have achieved a very high standard of all-round efficiency while in Germany, but Fallingbostl can hardly be described as a very popular garrison. There was some relief therefore when the Regiment moved to Northern Ireland at the end of 1971, despite the unrest in that unhappy province, and it replaced the 17th/21st Lancers at Omagh as part of the permanent garrison. It converted to the reconnaissance role and was equipped with armoured cars. Within only a few days of becoming operational the first casualty was suffered when Corporal Powell was mortally wounded by a mine while on patrol, and since then the Regiment has been constantly engaged on active operations. Aid to the Civil Power is the most distasteful of all a soldier's duties and the distaste is thrice compounded when the duties take place in one's own country. No

---

* The first phase of the Radfan campaign was commanded by Brigadier Lunt who had commanded the Regiment from 1957–59. The Tank Troop supporting his Force was commanded by his son, Second Lieutenant Lunt. This led to some wags in the Regiment describing the operations as 'The Lunts' Front'.

soldier can possibly enjoy operating against his own kith and kin, however misguided their motives or brutal their actions, but the job has to be done and the Regiment is going about it as it has always done, calmly, efficiently, and in the hope that wiser counsels will eventually prevail.

* * * *

What makes a 'good' regiment? All regiments consider themselves to be equally good, but as among George Orwell's pigs, some are more 'equally good' than others. Money used to have something to do with it. Expensive regiments tended to attract wealthy officers who could afford to set the pace in days when dress and drill were the main criteria of military efficiency. They became 'crack' regiments, but from the beginning of this century war has become much more complicated and military uniform much more utilitarian. There has also been a great levelling-out of wealth. Officers were once expected to purchase their various steps in rank; in 1840 the colonelcy of the 16th Lancers was valued at £14,000 (much more, of course, by today's monetary values) while a captain would have had to find £5,000 or more to purchase a troop. Officers without private means seldom rose above major. Purchase was abolished in 1872 but many regiments continued thereafter to require their officers to possess private means, particularly in the cavalry. Such conditions no longer prevail today. Mess life, once so expensive, has ceased to be a financial drain, and although uniform costs money, there is much less of it. Above all, modern weapons require highly-skilled men to handle them. There is no longer a place in the army for the amateur, although some may regret his passing from the scene on the grounds that he did lend tone to the vulgar business of war! The British Army today is the most professional army this country has ever before possessed in peace-time and all those serving in it are rightly proud of their high professional standards.

Regiments are judged by their efficiency nowadays and this embraces the entire spectrum of soldiering. The modern young man has little use for 'square-bashing' and has to feel that he is doing a worth-while job. He is better educated than his forefathers, more aware of the world about him, and disinclined to be treated as an automaton. He still has pride in his regiment, which for most young soldiers is the limit of their military horizon, but he has no use for the officer or N.C.O. who does not know his job. On occasions he has to carry out his duties in front of television cameras and display restraint in the face of intense and unmerited provocation. He has to handle immensely complicated machines. He may not be quite so tough as his forebears since he comes from a different environment but he is as capable as they were of feats of remarkable gallantry. He is probably more ambitious and less inclined to stay in the same old groove. If he knows his job thoroughly and finds satisfaction from doing it; if he respects his officers as equally competent professionals and they respect him; if he believes that he and his family are receiving a square deal from the Army and the Nation – then he will be a 'good' soldier and the regiment in which he serves will be a 'good' regiment.

This is what the Regiment has striven for throughout the past 25 years. If it has succeeded, it has been helped by two significant facts. Firstly, the Regiment has long been a 'Family' one. Although cavalry regiments have only recently been linked with Counties, sons have been following fathers into the Regiment for generations past; the sons of five former commanding officers were serving in the Regiment in 1964. The link with Staffordshire, which started in 1958, now means that brother has followed brother into the 'Scarlet Lancers,' and the Regiment contains a high proportion of Staffordshire men who make excellent soldiers. Secondly, the Regiment has long contained a family feeling in quite another sense. John Burgoyne set the tone when he first raised the 16th Light Dragoons and exhorted his officers to cultivate a friendly relationship with their soldiers. This has continued down the years. The rest of the Army has long considered the cavalry to be 'stand-offish' but the Regiment has always sought to avoid such a label being attached to it. 'We got on particularly well with all the other regiments,' wrote John Luard of Bhurtpore. 'The 16th Lancers never walked near our cooking places nor spat on our food,' wrote Subedar Sita Ram of the 1st Afghan War, comparing the Regiment favourably with other European regiments. General Gough, writing of the 16th Lancers in India in

1897, says, 'There was no snobbery in the 16th. We got on well with all regiments alike. We had a good opinion of ourselves but took good care not to show it.' And so it has remained right down to the present day.

*   *   *   *

Thus ends a chronicle of close on three hundred years. It is the story of two cavalry regiments which came together to make one regiment. The story began with horses and it ends with armoured cars and scout cars. Much else has changed in the course of three centuries, but one thing has not changed. To the reader of this book it may seem an odd quirk of fate that it has ended where it first began – in Northern Ireland, where old quarrels die hard.

Appendices

Appendix I

BATTLE HONOURS

At the time of their amalgamation with the 5th Royal Irish Lancers the 16th The Queen's Lancers had more Battle Honours than any other cavalry regiment of the line.

16th The Queen's Lancers

Beaumont, Willems, Talavera, Fuentes d'Onor, Salamanca, Vittoria, Nive, Peninsula, Waterloo, Bhurtpore, Ghuznee 1839, Afghanistan 1839, Maharajpore, Aliwal, Sobraon, Relief of Kimberley, Paardeberg, South Africa 1900-02, Mons, Le Cateau, Retreat from Mons, Marne 1914, Aisne 1914, Messines 1914, Ypres 1914, '15, Bellewaarde, Cambrai 1917, Somme 1918, Pursuit to Mons.

5th Royal Irish Lancers

Blenheim, Ramillies, Oudenarde, Malplaquet, Suakin 1885, Defence of Ladysmith, South Africa 1899-1902, Mons, Le Cateau, Retreat from Mons, Marne 1914, Aise 1914, Messines 1914, Ypres 1914, '15, Bellewaarde, Arras 1917, Cambrai 1917, Somme 1918, St. Quentin, Pursuit to Mons.

16th/5th The Queen's Royal Lancers

Fondouk, Bordj, Djebel Kournine, Tunis, North Africa 1942-43, Cassino II, Liri Valley, Advance to Florence, Argenta Gap, Italy 1944-45.

THE VICTORIA CROSSES OF THE REGIMENT

The Tirah Expedition, 1897
Lieutenant The Viscount Fincastle (later the Earl of Dunmore), 16th The Queen's Lancers.

South Africa, 1901
Lieutenant F. B. Dugdale, 5th Royal Irish Lancers.

France, 1917
Private Clare, 5th Royal Irish Lancers.

Appendix II

THE REGIMENTAL MARCH

The Regimental March, SCARLET AND GREEN, was arranged by Bandmaster T. Noble in 1950. The first four bars is the trumpet call of the 16th Lancers followed by the principal theme of the 16th Lancers quick march ENGLISH PATROL. The introduction to the trio is the trumpet call of the 5th Lancers which is followed by ST. PATRICK'S DAY which was the quick march of the 5th Lancers. The Grandioso is a combination of both the 16th and 5th Lancers marches.

The 16th Lancers slow march was also ENGLISH PATROL. The 5th Lancers slow march was LET ERIN REMEMBER followed by THE HARP THAT ONCE THROUGH TARA'S HALLS. The 16th/5th The Queen's Royal Lancers slow march is THE QUEEN CHARLOTTE, also arranged by Bandmaster T. Noble.

The Spanish National Anthem is always played after Regimental dinner nights in the Officers Mess immediately before the National Anthem.

In 1945 Bandmaster T. Noble introduced a Male Voice Choir and it has become a tradition for this choir to sing in the officers mess on regimental dinner nights.

Appendix III

REGIMENTAL CUSTOMS

1. Mention has already been made of the practice of crimping the lance pennons (p. 55), and of the probably apocryphal reason given for wearing the Sam Browne cross strap the wrong way round (p. 61). So far as the latter is concerned, the official reason given for this custom is that the 16th Lancers had always worn their belts in this fashion since they were introduced, and after amalgamation 5th Lancers officers adopted the same fashion.
2. In common with other cavalry regiments officers do not address their seniors in the Regiment as "Sir" except on parade. At all other times Christian names are used, but the commanding officer is always addressed as "Colonel".
3. There is no record of the Loyal Toast ever having been drunk by the 16th Lancers except at the Regimental Dinner and that of the Old Comrades' Association. In the 5th Lancers the Loyal Toast was not drunk before 1888 except at the Regimental Dinner. But after the 5th Lancers went to India the toast was drunk on guest nights and on official occasions. After amalgamation the 16th Lancers practice was followed. The reasons given for not drinking the Loyal Toast is that the Regiment's loyalty is beyond question, and because of the Regiment's connection with Queen Charlotte, and their affection for her, they refused to drink the health of the Sovereign, King George III.
4. D Battery, Royal Horse Artillery, supported the 16th Lancers at Mons in 1914, and on many occasions thereafter. The affiliation was so close that D Battery was always referred to as "Our Battery", while the Horse Gunners always referred to the 16th as "Our Regiment". After the Great War the Regiment presented D Battery with a mounted Lancer in silver, and D Battery reciprocated by presenting the Regiment with a model of the 13-pr. Gun in silver. The former is inscribed "To Our Battery", and the latter "To Our Regiment", and the association is treasured to this day.

Appendix IV

THE STAFFORDSHIRE YEOMANRY (QUEEN'S OWN ROYAL REGIMENT)

It was decided in 1947 to affiliate Yeomanry Regiments to Regular Regiments of the Royal Armoured Corps. Thus began a long and valued association between the Regiment and the Staffordshire Yeomanry whose headquarters at that time were at Burton-on-Trent.

The Staffordshire Yeomanry were raised in 1794 at the Swan Inn at Stafford. It fought as a horsed regiment in Palestine during the Great War, where it particularly distinguished itself, and was mobilized in 1939 as part of the 1st Cavalry Division, then commanded by Major-General J. G. W. Clark, himself a 16th/5th Lancer. It was horsed at the time but in 1942 it was mechanized. It took part in the battle of Alam Halfa in September, 1942, and later, as part of 8th Armoured Brigade, played a distinguished part in breaking through Rommel's defences at El Alamein. The regiment was present at most of the important engagements during the Eighth Army's advance across North Africa and earned itself a magnificent reputation. It was fortunate to be commanded for much of the time by Lieutenant-Colonel J. A. Eadie, a born soldier and a splendid leader of men. He later became Honorary Colonel of the Regiment. The Staffordshire Yeomanry returned from North Africa to take part in the Normandy Landings, where they were equipped with D.D. Tanks, and subsequently in the bitter fighting for the assault across the Scheldt and the crossing of the Rhine. No other Yeomanry regiment acquired a better reputation for itself during the Second World War than the Staffordshire Yeomanry and it was present at several of the most famous battles of that war.

For 20 years the Regiment provided the Staffordshire Yeomanry with its permanent staff, including for most of that time the Training Major, Adjutant, Quartermaster, and R.S.M. It was a connection which benefitted both regiments and was greatly valued in the 16th/5th Lancers. In particular it helped the Regiment to cement its association with Staffordshire from which county today a high proportion of its soldiers are drawn – and first-class soldiers they are too. It therefore came as a great shock to learn in 1967 that the Staffordshire Yeomanry's role (that of an armoured regiment) was to be changed to a Home Defence task, with a consequent drastic reduction in the numbers of the permanent staff. It was an even greater shock at the end of 1968 when the Staffordshire Yeomanry was reduced to a cadre of a handful of officers and men, virtually ending a history which had begun in 1794. All this was of course connected with the reorganisation of the Territorial Army, with its roots deep in Britain's past, but if there was any rhyme or reason for the disappearance of Yeomanry regiments with great traditions, it was lost on the Staffordshire Yeomanry. It was also a matter of deep regret for the Regiment, although the close connection with Staffordshire has continued, and even deepened. The 16th/5th The Queen's Royal Lancers are now as much a Staffordshire regiment as the Staffordshire Regiment itself.

In 1970 there was a change of policy, due to a change of government. An increase was authorised in the strength of the Territorial Army (which now has another name but which means much the same thing), and in consequence the Staffordshire Yeomanry was reborn in the shape of a Squadron of the Mercian Yeomanry, based on Stafford. The role is essentially infantry, if of a somewhat nebulous character, but the spirit of the Yeomanry remains. The first Squadron commander of the revived Staffordshire Yeomanry (Major Waring) was originally a 16th/5th Lancer, as was indeed the Yeomanry's last-but-one commanding officer (Lieutenant-Colonel Clutterbuck).

The Regiment is also affiliated with the 12th/16th Hunter River Lancers of the Australian Military Forces, and the 3rd Armoured Regiment of the Royal New Zealand Armoured Corps.